T0348578

VOLUME NINETY

Advances in
PROTEIN CHEMISTRY AND
STRUCTURAL BIOLOGY
Organisation of Chromosomes

VOLUME NINETY

Advances in
PROTEIN CHEMISTRY AND STRUCTURAL BIOLOGY
Organisation of Chromosomes

Edited by

ROSSEN DONEV

Institute of Life Sciences, College of Medicine
Swansea University, United Kingdom
Clinic and Polyclinic for Psychiatry and Psychotherapy
Rostock University, Germany

ELSEVIER

AMSTERDAM • BOSTON • HEIDELBERG • LONDON
NEW YORK • OXFORD • PARIS • SAN DIEGO
SAN FRANCISCO • SINGAPORE • SYDNEY • TOKYO
Academic Press is an imprint of Elsevier

Academic Press is an imprint of Elsevier
The Boulevard, Langford Lane, Kidlington, Oxford, OX5 1GB, UK
32 Jamestown Road, London NW1 7BY, UK
Radarweg 29, PO Box 211, 1000 AE Amsterdam, The Netherlands
225 Wyman Street, Waltham, MA 02451, USA
525 B Street, Suite 1800, San Diego, CA 92101-4495, USA

First edition 2013

Notice
No responsibility is assumed by the publisher for any injury and/or damage to persons or
property as a matter of products liability, negligence or otherwise, or from any use or
operation of any methods, products, instructions or ideas contained in the material herein.
Because of rapid advances in the medical sciences, in particular, independent verification of
diagnoses and drug dosages should be made.

ISBN: 978-0-12-410523-2
ISSN: 1876-1623

For information on all Academic Press publications
visit our website at store.elsevier.com

Printed and bound by CPI Group (UK) Ltd, Croydon, CR0 4YY
Transferred to Digital Printing, 2013

CONTENTS

PREFACE

Chromosomes undergo structural reorganizations throughout the cell cycle to enable their phase-specific functions. These reorganizations are very complex and numerous. Thus, in this volume of the *Advances in Protein Chemistry and Structural Biology* dedicated to the organization of chromosomes, we discuss only a few aspects of chromosome and cytoskeleton structural alterations enabling chromosome rearrangements.

Chapters 1 and 2 discuss the role of tandemly repeated DNA sequences in chromosome organization and function. The tandemly organized highly repetitive satellite DNA is the main DNA component of centromeric/pericentromeric constitutive heterochromatin. For a very long time, these sequences have been considered as a "junk DNA." However, it is now established that each chromosome possesses a unique code made up of different large tandem repeats. Tandem repeat multiformity tunes the developed nuclei 3D pattern by sequential steps of associations. Some satellite DNAs are also found transcribed during different phases of development and cell cycle, suggesting their importance for chromosome reorganization.

In Chapters 3 and 4, detailed reviews on the role of histone variants and zinc-finger proteins in elucidation of the structural dynamics of nucleosomes and chromosomes are presented. The presence of variant histone proteins in the nucleosome core raises the functional diversity of the nucleosomes in gene regulation and has the profound epigenetic consequences of great importance for understanding the fundamental issues such as the assembly of variant nucleosomes and chromatin remodeling. Zinc-finger DNA-binding proteins have shown to be important organizers of the 3D structure of chromosomes and as such are called the master weaver of the genome.

Chapter 5 focuses on the latest developments in unconventional actin configurations. The existence of actin in the nucleus was first reported more than 40 years ago. However, the idea of a nuclear actin met with skepticism for decades. Nuclear preparations were thought to be contaminated with actin from the cytoplasm. However, after the identification of actin in several nuclear complexes, which implicates it in diverse nuclear activities including transcription, chromosome remodeling, and nucleocytoplasmic trafficking, skepticism has finally given way to the challenging search for specific states and/or unique conformations actin may adopt to fulfill its nuclear functions.

Finally, Chapter 6 presents our current understanding of the chromatin condensation during mitosis with particular attention to the major molecular players that trigger and maintain this condensation. Furthermore, the mechanisms that take care to ensure that mitotic chromosome condensation and the block of transcription during mitosis do not wipe out the cell identity are also discussed.

It is my sincere hope that this overview of the basic mechanisms involved in the regulation of the dynamic chromosome organization would inspire future translational studies focusing on the fine regulation of the dysregulated cell cycle in diseases such as cancer, Alzheimer's, etc.

ROSSEN DONEV

Institute of Life Sciences, College of Medicine

Swansea University, United Kingdom

Clinic and Polyclinic for Psychiatry and Psychotherapy

Rostock University, Germany

CHAPTER ONE

Large Tandem Repeats Make up the Chromosome Bar Code: A Hypothesis

Olga Podgornaya[*,†,1], **Ekaterina Gavrilova**[†], **Vera Stephanova**[*],
Sergey Demin[*], **Aleksey Komissarov**[*]
[*]Institute of Cytology RAS, St. Petersburg, Russia
[†]Faculty of Biology and Soil Sciences, St. Petersburg State University, St. Petersburg, Russia
[1]Corresponding author: e-mail address: opodg@yahoo.com

Contents

Abstract

Much of tandem repeats' functional nature in any genome remains enigmatic because there are only few tools available for dissecting and elucidating the functions of repeated DNA. The large tandem repeat arrays (satellite DNA) found in two mouse whole-genome shotgun assemblies were classified into 4 superfamilies, 8 families, and 62 subfamilies. With the simplified variant of chromosome positioning of different

Advances in Protein Chemistry and Structural Biology, Volume 90
ISSN 1876-1623
http://dx.doi.org/10.1016/B978-0-12-410523-2.00001-8

tandem repeats, we noticed the nonuniform distribution instead of the positions reported for mouse major and minor satellites. It is visible that each chromosome possesses a kind of unique code made up of different large tandem repeats. The reference genomes allow marking only internal tandem repeats, and even with such a limited data, the colored "bar code" made up of tandem repeats is visible. We suppose that tandem repeats bare the mechanism for chromosomes to recognize the regions to be associated. The associations, initially established via RNA, become fixed by histone modifications (the histone or chromatin code) and specific proteins. In such a way, associations, being at the beginning flexible and regulated, that is, adjustable, appear as irreversible and inheritable in cell generations. Tandem repeat multiformity tunes the developed nuclei 3D pattern by sequential steps of associations. Tandem repeats-based chromosome bar code could be the carrier of the genome structural information; that is, the order of precise tandem repeat association is the DNA morphogenetic program. Tandem repeats are the cores of the distinct 3D structures postulated in "gene gating" hypothesis.

1. INTRODUCTION

Tandemly repeated sequences (TRs) are common in higher eukaryotes and can comprise up to 10% of the genomes. Much of TRs' functional nature in any genome remains enigmatic because there are only few tools available for dissecting and elucidating the functions of repeated DNA (Blattes et al., 2006). It is supposed that a certain amount of constitutive heterochromatin is essential in multicellular organisms at two levels of organization: chromosomal and nuclear. At the chromosomal level, TRs can be of profound structural as well as evolutionary importance, as genomic regions with a high density of TRs (e.g., telomeric, centromeric, and heterochromatic regions) often have specific properties such as alternative DNA structure and packaging (Kobliakova, Zatsepina, Stefanova, Polyakov, & Kireev, 2005; Podgornaya, Voronin, Enukashvily, Matveev, & Lobov, 2003; Ushiki & Hoshi, 2008; Vogt, 1990).

At the nuclear level of organization, constitutive heterochromatin may help maintain the proper spatial relationships necessary for the efficient operation of the cell through the stages of mitosis and meiosis. The association of satellite (or other highly repetitive) DNA with constitutive heterochromatin is understandable, as it stresses the importance of the structural rather than transcriptional roles of these entities. In the interphase nucleus, satellite DNAs (satDNAs) have one property in common despite their species specificity, namely, heterochromatization (Yunis & Yasmineh, 1971). Heterochromatization appears to involve RNA interference-mediated chromatin modifications (Alleman et al., 2006; Martienssen, 2003). The RNAi pathway is required for the formation of heterochromatin and silencing of repetitive sequences in *Drosophila melanogaster* (Grewal & Elgin, 2007). In mammalian cells, an RNA component is required for the association of HP1 with pericentric heterochromatin (Lu & Gilbert, 2007; Maison et al., 2002; Muchardt et al., 2002). Moreover, the strand-specific burst in transcription of pericentric satellites is required for chromocenter formation and early mouse development. Specific expression dynamics of major satellite (MaSat) repeats, together with their strand-specific control, represent necessary mechanisms during a critical time window in preimplantation development that are of key importance to consolidate the maternal and to set up the paternal heterochromatic state at pericentric domains (Probst et al., 2010). Such an important and crucial finding is based on the known sequence of the mouse MaSat. Most of the other mouse TRs could not be tested in similar experiments being undescribed and unclassified.

The analysis of the TR content of genomes is important for understanding the evolution and organization of genomes as well as gene expression. The computation approaches to the TR analysis on the genome level gradually appear with the genome sequencing advanced (Alkan et al., 2007; Ames, Murphy, Helentjaris, Sun, & Chandler, 2008; Mayer, Leese, & Tollrian, 2010; Warburton et al., 2008). Current review is dedicated to the large TRs or satellite DNA.

2. SATELLITE DNA IN THE MOUSE GENOME

2.1. Satellite DNA and tandem repeats

SatDNAs were identified about 50 years ago as an additional, "satellite," fraction of genomic DNA during the equilibrium centrifugation in CsCl gradient (Kit, 1961, 1962). SatDNAs have been found in all the high eukaryotic species investigated by advanced centrifugation methods (Beridze, 1986). Reassociation kinetics tests show that satDNA is the highly repetitive sequence, built up to 10% of the total DNA (Britten & Kohne, 1968; Corneo, Ginelli, & Polli, 1967, 1970). Cytologically distinguished domains for satDNAs have been shown with mouse MaSat with initially determined centromeric localization (Pardue & Gall, 1970). Simple sequence satellites with monomers of 5–15 bp have been found with the same method of equilibrium centrifugation of human DNA (Gosden, Mitchell, Buckland, Clayton, & Evans, 1975). The invention of the restriction endonucleases allows finding other satDNA families. Main human alpha-satDNA has been found with *Eco*RI and *Xba*I restriction endonucleases (Jabs, Wolf, & Migeon, 1984; Manuelidis, 1976, 1978a,1978b). Long satDNA blocks of up to millions of base pairs built of relatively short, from several up to hundreds of base pairs, monomers tandemly arranged "head to tail." In due time, the term "satDNA" has been broadened to comprise nearly all tandemly organized sequences found including mini- and microsatellites, which are located in euchromatin. It causes some confusion, for mini- and microsatellites possess features quite different from classical satDNA (López-Flores & Garrido-Ramos, 2012; Vogt, 1990).

Tandemly repeated DNA makes up a significant portion of the mammal genome. However, there is nothing for satDNA to abut to or to be "satellite" of anything in the assembled genomes; the arrays of tandem repeats just continue the euchromatic gene-containing regions. Nowadays, the term "large tandem repeats" looks like more adequate and we use it in this text. The term "satDNA" could not totally be drawn away, for some precise large tandem repeats have it included in their names historically (e.g., mouse MaSat, human satellite 3 (HS3)). On the other hand, the bulk of information obtained about classical satDNA by scientific community is quite applicable to the large tandem repeats.

Sequences of satDNAs are different in different species, and the level of their evolutionary variability is high (Beridze, 1986; Podgornaia, Ostromyshenskiĭ, Kuznetsova, Matveev, & Komissarov, 2009). However, they have some common structural features (Fitzgerald, Dryden, Bronson, Williams, & Anderson, 1994; Lobov, Tsutsui, Mitchell, & Podgornaya, 2001; Martínez-Balbás et al., 1990), and their positions at the chromosomes are fixed. A great part of satDNAs of different families reside in the pericentromeric or subtelomeric chromosome regions. All the eukaryote centromeric sequences investigated up to now, with the exception of simple *Saccharomyces cerevisiae* centromeres, contain satDNA as the main component (Choo, 1997; Pezer, Brajković, Feliciello, & Ugarković, 2012).

2.2. Mouse genome as a model for the tandem repeat search

Tandem repeat content on genome level is well investigated in the human genome and shows a wide range of repeat sizes and organizations, ranging from microsatellites of a few base pairs to megasatellites of up to several kilobases. Microsatellites and variable number of tandem repeats (called VNTRs or minisatellites) can be highly polymorphic and have an important use as genetic markers (Ames et al., 2008; Warburton et al., 2008).

The centromeric region of human chromosomes contains alpha-satDNA, the largest TR family in the human genome. This has been extensively studied and provides a paradigm for understanding the genomic organization of TRs (Schueler, Higgins, Rudd, Gustashaw, & Willard, 2001). Their arrays are composed of either diverged monomers with no detectable higher-order structure or as chromosome-specific higher-order repeat (HOR) units characterized by distinct repeating linear arrangements of an integral set of basic monomers (Rudd & Willard, 2004). The HOR structure of human centromeric alpha-satellite has been implicated as important in centromere function (Schueler et al., 2001).

In humans, the pericentromeric regions consist of alpha-satDNA arrays that are surrounded by arrays of "classical" satellites (e.g., human satDNA 1–4) (Choo, 1997; Lee, Wevrick, Fisher, Ferguson-Smith, & Lin, 1997; Moyzis et al., 1987; Podgornaya, Bugaeva, Voronin, Gilson, & Mitchell, 2000; Prosser, Frommer, Paul, & Vincent, 1986). These pericentromeric regions have a specific high-order chromatin structure and might be responsible for chromatin spatial organization.

In the house mouse, *Mus musculus*, there are two highly conserved TRs known as centromeric minor and pericentromeric major satellites (MiSat and MaSat, respectively, GSAT_MM and SATMIN in Repbase nomenclature). MiSat comprises an AT-rich, 120 bp monomer that occupies 300–600 kb of the terminal region of all mouse telocentric (single-armed) chromosomes and is the site of kinetochore formation and spindle microtubule attachment (Kalitsis, Griffiths, & Choo, 2006; Kipling, Ackford, Taylor, & Cooke, 1991; Wong & Rattner, 1988). MaSat (234 bp) is more abundant and resides adjacent to MiSat, and has a role in heterochromatin formation and sister chromatid cohesion (Broccoli, Miller, & Miller, 1990; Broccoli, Trevor, Miller, & Miller, 1991; Guenatri, Bailly, Maison, & Almouzni, 2004). Neither of these satDNA has been identified at the centromere of the morphologically distinct acrocentric Y chromosome (Hörz & Altenburger, 1981), which has a very small short arm that distinguishes it from the telocentric autosomes and X chromosome. Recently, the chromosome Y centromere was shown to comprise a highly diverged MiSat-like sequence (designated Ymin) with HOR organization previously not described for mouse MiSat arrays (Pertile, Graham, Choo, & Kalitsis, 2009).

2.3. Lack of tandem repeats in the reference genome

The large regions of classical heterochromatin are poorly covered by assembled sequences (Warburton et al., 2008), and for the mouse genome even less is known. Mouse acrocentric chromosomes have prolonged TR arrays at the ends, and it is the reason why these regions are difficult to assemble.

Table 1.1 Large TRs in the region adjusted to centromeric gap

Chromosome	TR subfamily	Array length (kb)	Coordinates (bp)
3	TRPC-21A-MM	33.6	3,000,001–3,033,629
4	TRPC-21A-MM	7.0	3,006,469–3,013,522
4	TR-22A-MM	4.9	3,104,899–3,109,811
6	TR-22A-MM	9.9	3,082,006–3,091,879
9	MaSat	38.4	3,000,003–3,038,419
11	MaSat	3.9	3,000,004–3,003,872
16	TRPC-21A-MM	9.0	3,232,335–3,241,336
17	TRPC-21A-MM	32.5	3,006,399–3,038,945
17	TR-27A-MM	4.6	3,070,530–3,075,093
18	TR-22A-MM	8.0	3,112,790–3,120,776

Only TRs with the array more than 3 kb in the distance up to 2 Mb from the centromeric gap are shown. Coordinates, the array position on chromosomes.

The mouse chromosomes end abruptly in 3 Mb gaps reserved for centromeric regions. We found out that only eight chromosome ends contain distinct TR arrays and only two of them contain MaSat (Table 1.1). The result obviously illustrated the bad assembly of the heterochromatic regions: the rest of the chromosomes contain genes at the ends. The assemblies did not cover even the beginning of the regions of the constitutive heterochromatin (Komissarov, Gavrilova, Demin, Ishov, & Podgornaya, 2011).

The sequences from those regions can be found in whole-genome shotgun (WGS) contigs. The advantage of WGS is that they include the entire shotgun sequencing reads, assembled into contigs. Both assemblies represent euchromatic and heterochromatic regions, even when not assembled into continuous contigs and/or not anchored on chromosomes yet. The regions enriched in TRs are mostly not anchored, although TRs and, in particular, large TRs are present in WGS due to their abundance in the genome.

2.4. Large tandem repeats in WGS assemblies

In WGS assemblies, the amount of all TRs is less than the experimentally determined amount of the MaSat alone (~8%) (Abdurashitov, Chernukhin, Gonchar, & Degtyarev, 2009), indicating that even in WGS assemblies TRs remain underrepresented (Table 1.2). All large TRs found in the mouse

Table 1.2 Tandem repeats in mouse WGS assemblies

Assembly	Size (bp)	Contigs	TRs (all)	% of assembly	TRs (>3 kb)
MGSC WGS	2,477,633,597	224,713	849,466	2.9	157
Celera WGS	3,003,109,157	837,963	1,084,552	5.0	784
Total WGS	5,480,742,754	1,062,676	1,944,018	3.8	941

TRs (all), total amount found in assembly; MSGC, the mouse sequencing genome consortium.

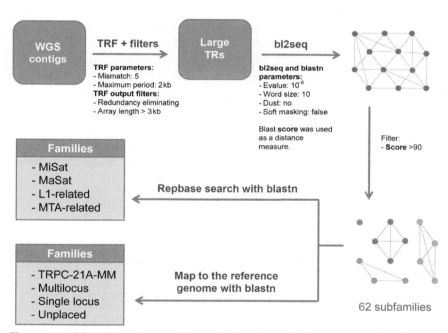

Figure 1.1 Schematic representation of the large tandem repeat classification workflow. Overview of the large tandem repeat analysis. For each program, only parameters that were changed are shown. Family names are given according to Table 1.2. The complete description of the workflow is given in Komissarov et al. (2011).

WGS were classified into 4 superfamilies, 8 families, and 62 subfamilies, including 60 not described yet, using classification based on array similarity, monomer length, the degree of unit similarity, position on the reference genome chromosome assemblies, and GC content. Each new subfamily was named according to the suggested cytogenetic-based nomenclature (the classification method shown on Fig. 1.1).

Among 62 subfamilies, only 2 subfamilies were similar with known mouse satDNAs from Repbase (database of repeated DNA elements): MaSat and MiSat (Table 1.3). The rest of the subfamilies are not present in Repbase and were therefore named according to their structure and genome position. For two families, the published nomenclature was used: single-locus family for arrays found only once in the reference genome, whereas multilocus family for arrays found at more than one locus (Ames et al., 2008). A subfamily name includes the letters TR (tandem repeat), genomic position (if known), minimal unit size in bp, index letter if there is more than one TR with similar unit size (A, B, etc.), and suffix MM (*M. Musculus*).

The characteristic feature of heterogeneous superfamily is the prominent variability of the TRs, which could be divided into subfamilies. The most abundant is TRPC-21A-MM, which has a strictly pericentromeric location. Multi- and single-locus families each represent ~6% of the TR dataset. Some of the TR arrays (~1%) from unplaced family, which have a distinct monomer and relatively long arrays, are not found in the reference genome. The next two families, MTA-related and L1-related (~3% together), show structural characteristics related to dispersed transposable elements (TEs), but were tandemly organized and have several features quite distinct from most members of the set (Table 1.3).

Table 1.3 Classification of large tandem repeat in the mouse WGS datasets

Superfamily	N	Family	Genome position	Arrays	% of TRs	Subfamilies
A. Centromeric	1	MiSat	Cen	21	2.2	1
B. Pericentromeric	2	MaSat	periCen	715	76.0	1
C. Heterogeneous	3	TRPC-21A-MM	periCen	50	5.3	1
	4	Multilocus	Any	57	6.0	20
	5	Single locus	Any	56	6.0	29
	6	Unplaced	Absent	11	1.2	8
D. TE-related	7	MTA-related	Any	15	1.6	1
	8	L1-related	Any	16	1.7	1

Eight families of the large TRs found in WGS were combined in four superfamilies (A–D). Families are formed according to sequence similarity and/or position in the reference genome.
Arrays, the number of TRs arrays found in WGS; % of TRs, percent of all found TRs in WGS; subfamilies, number of subfamilies in family; D, tandem repeats related to transposable elements (TE); MTA, mouse transcript retrotransposon, subfamily A; L1, L1_MM.

2.5. Large tandem repeat graphic representation

A monomer length, GC richness, and a degree of monomer diversity inside array are the most valuable parameters to distinguish and visualize TR arrays (Fig. 1.2A). The most clear and compact cloud is formed by MaSat arrays though it is not as monotonous as it might have been expected from the experimental studies (Vissel & Choo, 1989). MiSat cloud is in close proximity to the MaSat but forms a distinct group. In the area of relatively short monomer unit, clouds of TRPC-21A and other multilocus TRs are visible. The transposon-related TRs form a loose cloud in the region of long monomer units. Arrays from single-locus and unplaced families are scattered throughout the volume of the plot. Some of them are adjacent to the clouds of the well-formed subfamilies; this fact may reflect the underrepresentation of repeated DNA in genome sequences. It is likely that additional data from oncoming mouse genome sequencing could improve the classification of single-locus and unplaced families.

It is notable that most of the newly found subfamilies have GC content higher than MaSat and MiSat—the mean for TRPC-21A is ~50%, and even higher for multilocus family, ~57% (Fig. 1.2A). Both GC-rich and AT-rich satDNA are known in human and most of the high eukaryotes (Beridze, 1986; Palomeque & Lorite, 2008).

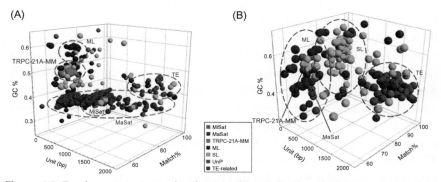

Figure 1.2 Tandem repeat array distribution. The graph of tandem repeat array distribution was done in Mathematica™ 7.0. Each circle represents one array found in WGS assemblies. Each family was colored according to Table 1.2 in different degrees of gray. X axis—monomer length (bp) up to 2 kb; Y axis—GC content is normalized to 1; Z axis—percentage of similarity between monomers inside array. (A) The graph of tandem repeat array distribution in the mouse WGS assemblies. Each circle represents one array. Main families' positions are indicated by circles. (B) The graph of tandem repeat arrays distribution in the reference genome. Main families' positions are indicated by circles. The MaSat and TRPC-21A-MM are marked by arrows.

There are significant changes in TR distribution, mainly due to the reduction of both pericentromeric MaSat and TRPC-21A in the reference genome (Fig. 1.2B). Multilocus and single-locus families form two groups; the third one is gravitate toward TE-related TRs. Multilocus group, with the core of TRPC-21A, extends along CG axis from 40% to 70% GC. Some representative of multilocus and single-locus arrays have a very long monomer of >1 kb. This makes them come visually into the area of the TE-related superfamily (Fig. 1.2B), although the structure of the long monomers does not show extensive similarity with known TE. However, TRs with long units that are classified as multilocus or single-locus families (based upon formal criteria) could actually be built on the base of very divergent or unknown TE.

The mouse reference genome has dramatic difference in large tandem repeat distribution in comparison with WGS assemblies, and graphic representation illustrates the assembly incompleteness (Fig. 1.2). The large TRs from WGS being recognized and annotated could improve assembling process.

2.6. TEs-related tandem repeats

Two families have structural similarity to TEs (Table 1.3). In both cases, an array is formed by a large monomer with a low degree of diversity and a similar GC content.

The first family, TR–MTA, is formed by MTA fragments: MTa, MaLR-LTR, and Mammalian apparent LTR–retrotransposons in Repbase. These LTR transposons have structural similarities to endogenous retroviruses, namely, ERV3, and are related to THE1 in humans (Smit, 1993). Endogenous retroviruses by themselves comprise ~10% of the mouse genome (Stocking & Kozak, 2008). Over time, most MaLRs have diverged considerably from their consensus sequence, so their number is now estimated at 25–94,000 copies (Smit, 1993). Preliminary analysis has not yet revealed significant similarities of the putative product of MTA ORF to any protein present in the databanks. The residual part of the ORF is now determined as internal part in MTA Repbase consensus, and it is included in TR arrays (Fig. 1.3A).

The second family, L1-related TRs, included part of the ORF2 and 3'LTR (Fig. 1.3A). In order to map TE-related arrays to the reference genome, two rules were applied. First, a TR hit at a chromosome locus counts as positive only when the alignment length is more than 2850 bp (95% from the original TR array limit of 3 kb). Second, a hit is counted as a single when the distance between two hits is less than 150 bp (5%). These rules allowed us to reduce the redundancy for some of the TE-related TR

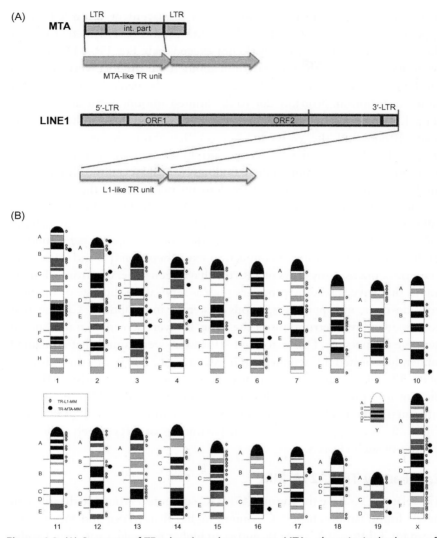

Figure 1.3 (A) Structure of TE-related tandem repeats. MTA—the principal scheme of MTA-related TR family; regions of MTA are denoted and the part from which TR unit arise is marked; LINE1—the principal scheme of L1-related TR family; L1 regions are denoted and the part from which TR unit arise is marked. (B) Chromosomal location of TE-related tandem repeats. Ideogram of mouse karyotype with MTA-like array positions indicated in black and L1-like array positions indicated in gray.

that came to the same locus with certain displacement discarded. After applying these rules, 284 hits with precise positions remained. Most of the loci found for TR–L1 family do not exceed 5 kb. For TR–MTA family, we found two loci with array length of about 10 kb. All loci were displayed on the banded chromosomes. There is no obvious regularity in TR–MTA family distribution, probably due to the limited amount of the arrays found (Fig. 1.3B, gray). The TR–L1 family definitely is enriched in heterochromatic bands, and the concentration on the chromosome X is visible (Fig. 1.3B, black). At the same time, no TE-related TR is found on the Y chromosome. Validation of these findings by FISH is technically challenging because the LTRs of other retroelements may obscure the results.

Heterochromatic regions are known to be enriched with LINE elements (Waterston et al., 2002). Although mammalian TEs represent almost half the DNA sequences in mammalian genomes, they are disproportionately understudied. Still, some intimate connection exists between TE and TRs. The genome contains many types of TRs that have been generated from TE fragments that have spread by processes such as replication slippage or unequal crossing-over, and appear as spurious clusters or interruptions. Clusters that contained many interruptions of the same TE were screened in the human genome for possible TR. A total of 40 clusters were found that contained tandemly repeated TE fragments (Giordano et al., 2007). So, the existence of the TE-related superfamily in the mouse genome is not surprising. But we did not find prolonged TR clusters based on TE-related TRs though the amount of such TRs is sufficient to form a distinct group (Table 1.3; Fig. 1.2). We suppose that different parameters of TRs search will allow finding of such clusters.

2.7. Family TRPC-21A-MM

TE-related TRs have the mean ~45% GC content. It is notable that most of newly found families have GC content higher than MaSat and MiSat—the mean for TRPC-21A is ~50%, and even higher for *multi/single-locus* family, ~57% (Fig. 1.2). Both GC-rich and AT-rich satDNA are known in almost all genomes of the high eukaryotes; hence our finding cures the strange asymmetric satDNA distribution reported for mouse until now.

In human genome, GC-rich TRs are represented by big classical satDNA, which are well recognized and studied extensively. Human satDNA 1–4 (HS 1–4) are based on a "simple" 5–6 bp motif and HS3 is mostly well investigated. Mouse TRPC-21A resembles human "classical" or "simple sequence" satellites in most features. These types of TRs have not been reported for mouse until now. In TRPC-21A, monomer units of ~5–7 bp could be distinguished inside the basic 21 bp monomer. Monomer 21 bp forms complex HOR structures, although the degree of diversity demands a special investigation to determine the exact oligonucleotide sequence. All the features of TRPC-21A are those of a "big classical" satDNA such as HS 1–4 (Richard, Kerrest, & Dujon, 2008). They are known to be chromosome specific. For example, the bulk of HS3 is located on chromosome 1, but it could be distinguished from HS3 on chromosome 9 (Enukashvily, Donev, Waisertreiger, & Podgornaya, 2007).

3. TANDEM REPEAT POSITION DEFINED BY FISH

3.1. Chromosomes of bone marrow cells: first set of probes

The aim of this part of the work was to check whether the bioinformatics predictions about the positions of newly found TRs could be supported by *in situ* experiments. We did not expect to obtain *in situ* the full correspondence of TR positions found *in silico*, as the heterochromatic part of the reference genome is far of being complete and TRs from all families are present in Chromosome Unknown. Nevertheless, *in silico* chromosome locations should be included in the set of the *in situ*-labeled chromosomes. At the first step, monomer units from three classes were selected for probe production. All probe sequences, probe design, and short description could be found in the original article (Komissarov et al., 2011).

TRPC-21A has a predicted *in silico* pericentromeric location, so the additional chromosomes have to be labeled accordingly. Single-strand dimer labeled from both ends yielded signal on nine chromosomes: 3, 5, 6, 7, 8, 12, 16, 17, and Y. In each case except the Y, the label was at the pericentromeric regions. Four chromosomes predicted as TRPC-21A bearing are in the set of *in situ*-labeled chromosomes. Chromosome 4 has a short TRPC-21A array *in silico*, but it lacks any signal, probably due to the wrong assembly. Another discrepancy is the pericentromeric signal seen on chromosome 7, while *in silico* TRPC-21A mapped to the internal 7D1 band. Instead, the Y chromosome bares the internal signal, which is probably due to the unique repeat content of the sex chromosomes (Pertile et al., 2009). On all chromosome spreads, the main signal belongs to chromosome 3.

The HOR structure of TRPC-21A suggests that chromosome-specific variants may exist. The next probe was based on the three units from chromosome 3. The probe is a double-stranded ~150 bp sequence of

TRPC-21A with ~20 bp flanking sequence that differs from the basic 21 bp unit. Flanks give the possibility to label probe alongside by PCR. This probe hybridizes to two chromosomes (3 and 17), in concordance with the large TRPC-21A fields at the ends of these chromosomes (Table 1.1). Thus, probes based on a long TRPC-21A fragment show specificity to just two chromosomes. It is possible that probes designed on the basis of variants of TRPC-21A that are seen *in silico* could be specific for other chromosomes.

TR-22A-based monomeric single-strand probe labeled from both ends hybridized to 10 chromosomes, 4 of them predicted as TR-22A bearing. In this case, the majority of the signal is seen at the pericentromeric regions (chromosomes 2, 6, 7, 9, 11, 17, 18), with additional sites of labeling seen on the arms of chromosomes 2 and 15, and in the subtelomeric region of chromosome 13. In each case, signal was located in heterochromatic dark bands. The permanent cell line L929 is known to have chromosome rearrangements (Kuznetsova, Voronin, & Podgornaya, 2006; Mamaeva, 2002), and the amount of signal is higher on L929 chromosome spreads comparing with those from cells from normal bone marrow, suggesting that the rearrangement is within heterochromatic regions. There is no obvious main signal on any chromosome spread, so the construction of the chromosome-specific probe on the base of TR-22A could be more complicated and, moreover, the fields at the ends of chromosomes 4, 6, and 18 do not exceed 10 kb (Table 1.1).

The single-locus TR-54B was selected due to the abundance of its arrays found at the XA1.2 pericentromeric band. A double-strand dimer probe was designed and labeled by PCR. About half of the signals obtained in early prophase chromosome spreads belong to the long loops emerged from chromosome ends. The signal on the X chromosome is situated at the predicted region. However, this signal as well as most of the rest could only be recognized on "fuzzy" chromosomes, when all the DAPI-stained material is visible but bands are obscure. Such a signal was designated as belonging to the long loops. Long loops are known to emerge from chromosomes from subtelomeric region during inevitable osmotic shock, which is a necessary step during chromosome-spread isolation (Kireeva, Lakonishok, Kireev, Hirano, & Belmont, 2004; Ushiki & Hoshi, 2008). The TR-54B location needs more investigation, but we supposed that it is representative of the subtelomeric heterochromatic field. Whether TR-54B is situated mostly in subtelomeric region should be checked with double FISH with probe attributed to this region (Kalitsis et al., 2006). In contrast to the bioinformatics predictions, TR-54B is not a single-locus TR because about 50 signals in total are visible on chromosome spreads (Table 1.4).

Table 1.4 Chromosome positions of different TRs in mouse genome from the data of all FISH experiments. (For interpretation of the references to color in this table legend, the reader is referred to the online version of this chapter.)

## chrms	TRPC-21A	TR-22A	TR-54B	TR-31A	TR-31B	TR-38C	TR-24B
1			cen_intr_tel				tel/–
2		intr_tel		cen/ tel		cen/–	cen/–
3	cen					cen_tel/tel	
4		intr	cen_tel	intr_tel/cen_tel	intr_tel	tel	tel/–
5	cen			cen/cen_tel	cen_intr_tel/cen_tel	tel/–	intr_tel
6	cen	cen		cen_tel / tel	tel		
7	cen	cen	cen_intr_tel	intr/ cen_intr	cen_intr_tel	–/intr	tel/–
8	cen			cen_intr	cen_intr/intr	intr	intr/ cen_intr
9		cen	cen_tel	tel	tel	cen_tel	cen/–
10			cen_tel	cen_intr		cen_tel / tel	
11		cen	intr_tel	cen_intr	cen	tel	cen_intr_tel
12	cen				cen/–	cen_intr/intr_tel	cen
13		tel	cen_intr_tel	intr/–	cen_tel	cen_tel / tel	
14			cen_intr_tel	cen / cen_ intr	cen		intr/cen_intr
15		intr			intr	cen/–	
16	cen			intr/–	tel		
17	cen	cen	intr_tel	cen	tel	cen_intr/intr	cen
18		cen	cen_intr	intr/tel	intr_tel	tel/–	cen_intr/intr
19			cen_tel			cen	intr/
X			cen	tel/ cen_tel	tel	tel/ intr_tel	cen_intr_tel
Y	intr	–(♂)	cen	–(♂)/–(♂)	–(♂)	intr	– (all ♀♀)

TR name indicated at the top, chromosome Ns—first column. cen, pericentromeric; tel, subtelomeric; intr, interstitial position on the chromosome. Reproduced positions are colored; positions observed on some metaphase plates indicated, but not colored.

3.2. Chromosomes of embryonic fibroblasts: second set of probes

Chromosome positions in the second set of probes were checked on metaphase spreads from mouse embryonic fibroblasts. The primary fibroblast culture was established from the skin of a newborn mouse. The fibroblasts were isolated by enzymatic digestion of skin sections (Solovjeva, Demin, Pleskach, Kuznetsova, & Svetlova, 2012). The procedure supposed that the cells used for mapping had normal karyotype without any chromosomal rearrangement.

The choice of subfamily to check by FISH was based on the unit size convenient for oligonucleotide production (20–40 bp), TRs' presence in the Chromosome Unknown (artificial chromosome with unplaced contigs), and bioinformatics information about position in the reference genome. All the probes were short—mostly common monomer, both ends labeled. Four probes gave good FISH results (Table 1.4). For each subfamily, several plates were karyotyped and signals on chromosomes marked according to combined results.

All the chromosomes expected are labeled. If in the first set of probes only TR-54B gave subtelomeric signal, then in the second set subtelomeric signals are prevailing.

Telomeres are involved in chromocenter formation (Dmitriev et al., 2002; Markova, Donev, Patriotis, & Djondjurov, 1994). Probably, the association is going on via subtelomeric large TRs, but the supposition could not be checked due to the probes' absence: subtelomeric region is far off from being complete. We conducted association analysis of centromeric minor MiSat and pericentromeric MaSat in the WGS assemblies. The monotonous MiSat and MaSat arrays as well as MiSat transition to MaSat were found, whereas not a single transition to a telomeric repeat was observed. Moreover, large telomeric repeats are totally absent in WGS assemblies (Komissarov, Kuznetsova, & Podgornaya, 2010). That means that heterochromatic TRs-enriched long fields prevent the reads combining in contigs around telomere.

Only several mouse centromere–telomere-containing regions of the chromosomes deserved special efforts to obtain a detailed sequence (Kalitsis et al., 2006). MiSat is counted to be centromeric, and conservative $(TTAGGG)_n$ is determined as telomere (Kipling, 1995). However, what remains unknown are the precise DNA sequence and molecular organization of the mouse telocentric p-arms or, indeed, those of other eukaryotic telocentric chromosomes. Several attempts at isolating the p-arm telomeric and subtelomeric sequences of the mouse chromosomes have largely been unsuccessful because of the repeating nature of the DNA residing in these regions and difficulties associated with anchoring these sequences to individual chromosomes. The authors (Kalitsis et al., 2006) have identified fosmid clones generated for the C57BL_6 mouse genome project, containing large randomly sheared inserts that bridge the gap between the p-arm telomere and centromere. The large-insert fosmid clones that span the telomere and centromere of several mouse chromosome ends have been identified.

A telomeric tandem repeat and a MiSat monomer unit were used to BLAST search the end-sequenced library. Twelve fosmid clones were identified with both telomere and MiSat hits, suggesting that the telomere and centromere were within 40 kb. Sequence analysis shows that the distance between the telomeric $(TTAGGG)_n$ and centromeric MiSat ranges from 1.8 to 11 kb. The telocentric regions of different mouse chromosomes comprise a contiguous linear order of $(TTAGGG)_n$ repeats, a highly conserved truncated long interspersed nucleotide element 1 repeat (tL1), and varying amounts of recently discovered telocentric TRs that share considerable identity with, and is inverted relative to, the centromeric MiSat. Sequencing between the tL1 and the MiSat array revealed a repeat with a monomer unit of 146 bp in some of the fosmid clones. This repeat, which is called TLC, is found in three of the four possible groupings of chromosomal loci represented by the different telomere-MiSat-containing fosmids.

FISH on metaphase spreads of mouse normal cells using a 3.5-kb subclone of TLC probe under no cross-hybridizing conditions for MiSat have been performed to confirm that the TLC repeat was specific to the telocentric ends of the mouse chromosomes. Positive signals that colocalized with those for the MiSat were observed on most metaphase chromosomes (Kalitsis et al., 2006). Not one of the long q-arm end bears TLC signal. This result shows that authors successfully found new TR for short p-arm, but the rest of TRs which full up the subtelomeric region of long q-arm could not be fished out by such an approach.

The members of TE-related family of TRs described above (Table 1.2 and Fig. 1.3) are similar to tL1 as being tandemly repeated derivatives of truncated L1. We did not find repeats similar to TLC, probably due to its similarity to MiSat, which is underrepresented in WGS. Instead of that, we found a number of TRs, which mark the subtelomeric region of q-arm. This is not surprising, for none of the chromosome assemblies in the reference genome are continuous from last gene, that is, euchromatic region, up to the telomere. So the heterochromatic regions at the chromosome ends are lost in assemblies in the same way as in centromeric ones.

All the probes gave multiple signal in spite of probes having been selected from multilocus (2 hits) or unplaced (1 hit) family. The predicted arrays for TR-31A (7D1, 8C1) and TR-31B (7D1, 14A2) are of the order of ~5 kb, and they are visible as centromeric or internal signal. Compare to this signal size, the subtelomeric arrays of TRs could be very large—up to 50 kb. Centromeric and internal signals are an order weaker than huge array at q-arm.

3.3. Hybridization summary

Mouse chromosomes are defined telocentric chromosomes having no obvious short arm at a cytogenetic level. The Y chromosome contains a short p-arm and can be defined as acrocentric. Molecular cytogenetic studies have placed the mouse centromeric MiSat DNA 10–1000 kb away from the end of the telocentric short arm (p-arm) (Garagna, Zuccotti, Capanna, & Redi, 2002; Kipling et al., 1991; Narayanswami et al., 1992). So it is visible that cytogenetic resolution of conventional wide-field microscopy is rather coarse. MaSat and MiSat were recognized and cloned in the 1970s–1980s (Hörz & Zachau, 1977; Joseph, Mitchell, & Miller, 1989; Pietras et al., 1983; Wong & Rattner, 1988). It took about 20 years for them to become an established marker for centromere and pericentromere region. We found 8 families of TRs including 62 subfamilies and characterized them by bioinformatics analysis. Using WGS, we have identified about 200 *in silico* predicted TRs. We are only at the inception of the mapping process by the FISH experiments. Nevertheless, we combine the preliminary results of all FISH in Table 1.4.

The most probes used are the common monomers having both ends labeled. It means that all the arrays, not chromosome-specific variants, are recognized. Most of the TRs are found in the pericentromeric and subtelomeric (or both) regions as expected. In the cases where signal is found on some plates but absent in the rest, probably, it means that the array size is close to the limits of olygo-FISH resolution. The poor assembly of the heterochromatic regions in the reference genome does not allow the checking of this supposition.

Even a simplified variant of chromosome positioning of different TRs, without detection of chromoband, to which the signal belongs, makes it visible that each chromosome acquires a kind of unique "bar code" made up of different TRs.

We are only in the beginning of the construction of a physical map of TRs and have hardly tested most frequent ones when we noticed the non-uniform distribution instead of the positions reported for MaSat and MiSat. Most of the new TRs are more GC rich than both MaSat and MiSat. HOR structure was determined for many of them, suggesting the existence of TR chromosome-specific variants. The chromosome-specific variant proved to exist for TRPC-21A. Bioinformatics prediction tells us the MaSat can possess chromosome-specific variants and probes to check it could be designed and even MiSat could be not as uniform as it would supposed earlier. So the amount of variable TRs in the genome altogether with their ability to be relatives (to belong to definite families) and, on the other hand, the diversity as chromosome-specific variants make TRs good candidates to build up individual unique bar code for each mouse chromosome.

4. "BAR CODE" HYPOTHESIS

4.1. Internal large tandem repeats in human and mouse reference genomes

Alpha-satDNA is the only functional DNA sequence associated with all naturally occurring human centromeres. Two distinct forms of an alpha-satellite are recognized based on their organization and sequence properties. A large fraction of an alpha-satellite is arranged into HOR arrays where corresponding monomers are organized as multimeric repeat units ranging in size from 3 to 5 Mb (Mahtani & Willard, 1990; Paar et al., 2005). While individual alpha-satellite units show 20–40% single-nucleotide variation, the sequence divergence between HOR units is typically less than 2% (Warburton et al., 2008). The organization and unit of periodicity of these arrays are specific to each human chromosome, with the individual monomer units classified into one of five different suprafamilies based on their sequence properties (Lee et al., 1997). Human chromosome-specific probes based on alpha-satDNA (Alexandrov, Kazakov, Tumeneva, Shepelev, & Yurov, 2001), "classical" satDNA (Enukashvily et al., 2007), and megasatellites (Warburton et al., 2008) exist and are used in cytogenetic analysis. It appears that using human WGS and assembled genome, a set of TRs characteristic for each human chromosome could be found, suggesting that TRs might provide a kind of "bar code" for individual chromosomes.

We tried to make TRs' bar code visible with bioinformatics approach, though only the euchromatic part of the genome and consequently only the internal heterochromatin could be analyzed in any reference genome database. All TRs with array size no less than 3 kb in the human and mouse reference genomes were mapped *in silico* to the reference chromosomes. In this case, the color of mapped TRs is determined only by GC content. Acrocentric mouse chromosomes lack pericentromeric and centromeric regions characterized by AT richness (black) and little is known about telomeric/subtelomeric region (Fig. 1.4, mouse). Reference genomes allow marking only internal TRs, and even with such a limited data, the colored bar code made up of TRs is visible.

We can suppose that code is even more obvious in currently unassembled pericentromeric and subtelomeric regions (Fig. 1.4, hypothetical). Namely, these regions are involved in chromocenter formation (mouse) and in chromosome association. They should bare the mechanism for chromosomes to recognize the regions to be associated. The hypothesis could not be checked without filling these blank regions with precise sequences.

The lack of mouse chromosome-specific probes causes problems for most genome-connected studies, including studies in developmental biology. Using WGS, we have identified a set of large tandemly repeated DNA. The next step is to map most of them to check whether there is the chromosome specificity in the hybridization pattern. Probably, it will be possible to create individual chromosome bar code set of probes to be used in cytogenetic analysis. We propose that this bar code is the heterochromatin signature of

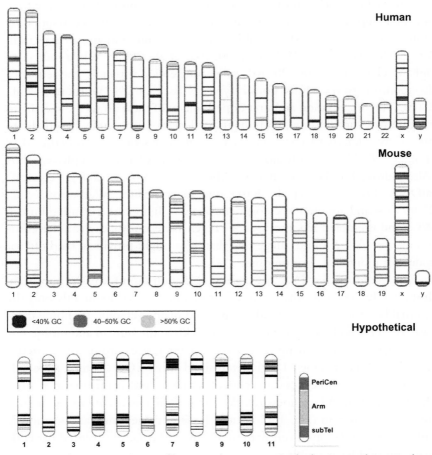

Figure 1.4 Schematic positions of large tandem repeats in the human and mouse chromosome of reference genomes. Tandem repeats are colored by GC content. According to "bar code" hypotheses, each chromosome could have unique set of large tandem repeats at pericentromeric and subtelomeric regions; arms are not shown (white space).

each chromosome, and these signatures help to arrange chromosomes in the nucleus in the specific order during development.

A given DNA sequence structure can express its potential genetic function in various ways. A genetic function to a DNA sequence is that its organizational pattern has a "code," that is, instructive for specific molecular (multimolecular) interactions or processes. The most investigated and well known is "triplet code" (Lengyel, Speyer, & Ochoa, 1961). The second code of the DNA sequence was based on nucleosome structure and positioning and named "chromatin code" (Trifonov & Sussman, 1980). The third one recognized underlines the "coding" potential of a tandem-repetitive DNA structure. The ability to adopt a specific folding structure of TR chromatin chain based on TR locus-specific super-repeat structure, stabilized and perhaps also triggered by the binding of specific nuclear proteins, becomes evident. TR-based code has been termed the "chromatin folding code," as an extension of the term "chromatin code" (Vogt, 1990). Thus, TRs acquire the function as the carrier of the genome structural information, but the mechanisms of the expression of this kind of information and its role in the genome housekeeping are still obscure.

4.2. Proteins specific in binding to tandem repeats

Relative progress has been achieved in finding the proteins, specific in binding to TRs. The features of these proteins give a clue for the beginning of comprehension of chromatin 3D organization. Telomere-binding TRF2/MTBP attaches telomeres to the membrane in the interphase due to its rod-domain-like motif (Podgornaya, Bugaeva, et al., 2000; Voronin, Lobov, Bugaeva, Parfenov, & Podgornaia, 1999a,1999b). Both p68—DEAD/RNA-helicase (Enukashvily, Donev, Sheer, & Podgornaya, 2005) and SAF-A/hnRNP-U (Lobov, Tsutsui, Mitchell, & Podgornaya, 2000)—are satDNAs binding proteins that possess an ATPase domain. The dynamic pattern of their distribution confirms that proteins are able to form dynamic complexes of different overlapping compositions (Kukalev, Enukashvili & Podgornaia, 2005; Kukalev, Nord, Palmberg, Bergman, & Percipalle, 2005) in accordance with the highly dynamic nature of the nuclei (Phair & Misteli, 2000; Phair et al., 2004).

Interphase nuclei could be imagined as a number of sponge-like ruffle-round chromosome territories which could be quite unsteady at their surface. SAF-A/hnRNP-U and p68-helicase are examples of proteins suitable to be involved in precise compartment motion. Their location in the inter-chromosome territory space, ATPase, and coiled-coil domains and the ability to be bound by TR make them suitable to be part of the wires with the help of which TR (satDNAs) and subsequently the overall chromosome territory are rotating. In the case of active transcription, p68-helicase can be involved in local "gene expression matrices" formation (Enukashvily, Malashicheva, & Waisertreiger, 2009) and due to its satDNA binding specificity cause the local chromosome territory rearrangement. The marks of chromatin rearrangement which should be heritable during cell division could be provided by SAF-A/hnRNP-U (Kukalev, Lobov, Percipalle, & Podgornaya, 2009). During telophase unfolding, the proper chromatin arrangement got restored according to these marks. The structural specificity of binding of both proteins to the TR provides a regulative but relatively stable mode of binding. The structural specificity of protein binding could also help to find the "magic" centromeric sequence (Podgornaya, Dey, Lobov, & Enukashvili, 2000; Podgornaya et al., 2003).

The problem with TR-binding proteins is that their selectivity is not strong enough to provide the association of precise chromosomes' regions. For example, p68 helicase binds human alpha-satDNA (Enukashvily, Lobov, Kukalev, & Podgornaya, 2000) as well as MiSat (Enukashvily et al., 2005). Recently, RNA-mediated mechanism, which can provide higher selectivity of association, has been found.

4.3. RNA-mediated mechanism of tandem repeat association

The organization of interphase chromosomes into distinct units called "territories" is commonly accepted as well as the implications of this packaging for regulation of gene expression. It is suggested that the tethering of chromatin loops maintains the structural integrity of chromosome territories. The stable attachments are thought to provide structural integrity and be associated with nontranscribed DNA, that is, tandem repeats (Podgornaya et al., 2003).

Chromosome territory architecture appears to depend on ongoing transcription (Haaf & Ward, 1996). RNA polymerase II inhibition leads to satDNA blocks dissociation into extended strands and to chromosome territory disruption. This is the evidence of a deep connection between transcription and heterochromatization. The speed of rearrangement reflects the fact that heterochromatin compactness is dependent on transcription activity (Haaf & Ward, 1996).

It happens that TRs are not as totally silent as it has been expected. Transcription of pericentromeric satDNA is a general stress response in human cells (Valgardsdottir et al., 2008). We tried to test whether satDNAs can be involved in chromosome territories interaction. We studied the localization, level of condensation, methylation, and transcriptional status of centromeric (alphoid DNA) and pericentromeric (HS3) sequences throughout the cell cycle in different cell types. Alphoid DNA remained condensed and heavily methylated in all cell types. On the contrary, HS3-1 participates in heterologous chromosome, extends into the territory of neighboring chromosomes, and expresses in malignant and senescent cells. The level of HS3-1 decondensation was higher in aged primary fibroblasts as compared to malignant cell line. The HS3-1 transcripts are transcribed from the reverse chain and polyadenylated. Thus, the involvement of satDNA in associations between human chromosomes and intermingling of chromosome territories was demonstrated. The invading satDNA undergoes decondensation to a certain level. This process is accompanied by demethylation and transcription (Enukashvily et al., 2007). The satDNA transcription found in malignant (cancer) and senescent cells is the reflection of the normal process going on in development. Pericentromeric but not centromeric satDNA transcription accompanied the mouse embryonic stem cell induction into differentiation. In this case, RNA was polyadenylated and transcribed from the forward chain. In induced cells, pericentromeric satDNA interacts with the RNA-helicase p68 that was previously shown to be satDNA binding protein (Enukashvily et al., 2009).

A unique strand-specific regulation of MaSat repeats during mouse preimplantation development has been shown. The expression of MaSat as well as major changes in their subnuclear higher-order organization illustrates the extremely dynamic and deliberate processes that occur during narrow time window when maternal pericentric heterochromatin is reorganized, and the paternal one is set up for subsequent development (Probst et al., 2010).

The developmental arrest observed after the depletion of MaSat transcripts strongly supports the functional relevance of these transcripts. Either the act of transcription and the associated chromatin remodeling or the transcripts as a structural component or through their processing could be important. Indeed, the rapid disappearance of MaSat transcripts by the four-cell stage implies both transcriptional and posttranscriptional processing. In analogy to RNAi-mediated degradation and RNA-directed transcriptional silencing mechanisms in *S. pombe* (White & Allshire, 2008), it is tempting to speculate that either major transcripts could fold into partially double-stranded secondary structures (Djupedal & Ekwall, 2009) or the two complementary transcripts hybridize to double-stranded intermediates, which are further processed to trigger heterochromatin formation at the paternal domains.

Such a scenario is attractive. However, to determine whether small double-stranded RNAs corresponding to MaSat transcripts accumulate at the two-cell stage, a detailed analysis including deep sequencing of small RNA libraries derived from the corresponding developmental stages would be necessary. Alternatively, mechanisms independent of RNAi, as found in *S. cerevisiae*, during which antisense transcripts act *in cis* or *in trans* on homologous target sequences to induce transcriptional silencing (Camblong, Iglesias, Fickentscher, Dieppois, & Stutz, 2007), can also be envisaged. Future studies should elucidate the pathways involved in degradation as well as the possible interaction with mechanisms operating at the transcriptional level that together result in the low constitutive expression level at later developmental stages (Probst et al., 2010).

We suppose that this finding reflects the first step of the successive inclusion of different TRs into association. In principle, the realization of the morphogenetic TRs program will be as follows. The MaSat of each species is involved in initial paternal and maternal genomes association as it is shown for mouse. A variety of other TRs, probably, in order of its abundance are acting as marks for chromosomes and their regions' association in parallel to the main differentiation steps. The burst of precise TR transcription in precise time would signify the following association. Chromosome-specific TR variants will add to the fine selectivity of the association. The associations, initially established via RNA, become fixed by histone modifications (the histone or chromatin code) and specific proteins. In such a way, associations, being at the beginning flexible and regulated, that is, adjustable, appear as irreversible and inheritable in cell generations. TR multiformity tunes the developed nuclei 3D pattern by sequential steps of associations. On the other hand, TR species specificity reflects the unique morphology of the living organisms.

The scheme suggested needs a lot of work to clarify, but the key point is the absence of TR nomenclature and classification and consequent lack of knowledge about heterochromatin DNA. Our attempt based on mouse genome gives a clue for TR investigation in other genome databases.

4.4. "Bar code" as morphogenetic program

The implications from the "gene gating" hypothesis (Blobel, 1985), when expressed in "nucleographical" terms, postulated that a given expanded (i.e., active) gene in the G1 phase nucleus of a certain tissue cell is a specific reference coordinate on the surface of the nucleus analog, in geographical terms, of the city of Eisenach on the surface of the earth. All other compacted or expanded genes would then be 3D coordinated with respect to this reference point and acquire the position beneath the city of Salzburg or Bonn. The genome of a higher eukaryotic is organized into a number of distinct 3D structures, each characteristic of a given differentiated state. These discrete 3D structures are envisioned to develop in a hierarchical and largely irreversible manner from an omnipotent 3D structure of the zygote genome. The information for establishing the precise 3D structure is assumed to reside in the genome (Blobel, 1985). This is accepted in principle but molecular mechanisms of such an organization remain obscure. The molecular mechanisms of several key points have to be clarified in order to prove this hypothesis.

"Gene gating" hypothesis has been published about 30 years ago (Blobel, 1985). Now, some sound finding and implications could be added to the hypothesis. Centromeric and subtelomeric regions of constitutive heterochromatin are supposed to play an important role in chromatin positioning in the interphase nucleus (Manuelidis & Borden, 1988). TRs, which are essential components of these regions and internal heterochromatin, are not only able to acquire a unique structure (Vogt, 1990) but are of variable families and possess chromosome-specific variants (Fig. 1.2; Table 1.4). TRs' precise but regulative association is reached via RNA-mediated mechanism and then fixed by histones and TR-binding proteins. TR-based chromosome bar code becomes the carrier of the genome structural information, that is, the order of precise TR association is the DNA morphogenetic program. TRs are the cores of the distinct 3D structures postulated (Blobel, 1985).

ACKNOWLEDGMENTS

The work was supported by MCB grant from Presidium of Russian Academy of Sciences (N01200955639) and RFBR (N11-04-01700-a). We would like to thank Dr. N. Pleskach for the help with metaphase plate preparations.

REFERENCES

Abdurashitov, M. A., Chernukhin, V. A., Gonchar, D. A., & Degtyarev, S. K. (2009). GlaI digestion of mouse gamma-satellite DNA: Study of primary structure and ACGT sites methylation. *BMC Genomics, 10*, 322.

Alexandrov, I., Kazakov, A., Tumeneva, I., Shepelev, V., & Yurov, Y. (2001). Alpha-satellite DNA of primates: Old and new families. *Chromosoma, 110*(4), 253–266.

Alkan, C., Ventura, M., Archidiacono, N., Rocchi, M., Sahinalp, S. C., & Eichler, E. E. (2007). Organization and evolution of primate centromeric DNA from whole-genome shotgun sequence data. *PLoS Computational Biology, 3*(9), 1807–1818.

Alleman, M., Sidorenko, L., McGinnis, K., Seshadri, V., Dorweiler, J. E., White, J., et al. (2006). An RNA-dependent RNA polymerase is required for paramutation in maize. *Nature, 442*(7100), 295–298.

Ames, D., Murphy, N., Helentjaris, T., Sun, N., & Chandler, V. (2008). Comparative analyses of human single- and multilocus tandem repeats. *Genetics, 179*(3), 1693–1704.

Beridze, T. (1986). *Satellite DNA.* Berlin: Springer-Verlag, p. 149.

Blattes, R., Monod, C., Susbielle, G., Cuvier, O., Wu, J., Hsieh, T., et al. (2006). Displacement of D1, HP1 and topoisomerase II from satellite heterochromatin by a specific polyamide. *The EMBO Journal, 25*(11), 2397–2408.

Blobel, G. (1985). Gene gating: A hypothesis. *Proceedings of the National Academy of Sciences of the United States of America, 82*(24), 8527–8529.

Britten, R. J., & Kohne, D. E. (1968). Repeated sequences in DNA. Hundreds of thousands of copies of DNA sequences have been incorporated into the genomes of higher organisms. *Science, 161*(3841), 529–540.

Broccoli, D., Miller, O. J., & Miller, D. A. (1990). Relationship of mouse minor satellite DNA to centromere activity. *Cytogenetics and Cell Genetics, 54*(3–4), 182–186.

Broccoli, D., Trevor, K. T., Miller, O. J., & Miller, D. A. (1991). Isolation of a variant family of mouse minor satellite DNA that hybridizes preferentially to chromosome 4. *Genomics, 10*(1), 68–74.

Camblong, J., Iglesias, N., Fickentscher, C., Dieppois, G., & Stutz, F. (2007). Antisense RNA stabilization induces transcriptional gene silencing via histone deacetylation in S. cerevisiae. *Cell, 131*(4), 706–717.

Choo, K. H. (1997). Centromere DNA dynamics: Latent centromeres and neocentromere formation. *American Journal of Human Genetics, 61*(6), 1225–1233.

Corneo, G., Ginelli, E., & Polli, E. (1967). A satellite DNA isolated from human tissues. *Journal of Molecular Biology, 23*(3), 619–622.

Corneo, G., Ginelli, E., & Polli, E. (1970). Different satellite deoxyribonucleic acids of guinea pig and ox. *Biochemistry, 9*(7), 1565–1571.

Djupedal, I., & Ekwall, K. (2009). Epigenetics: Heterochromatin meets RNAi. *Cell Research, 19*(3), 282–295.

Dmitriev, P. V., Prusov, A. N., Petrov, A. V., Dontsova, O. A., Zatsepina, O. V., & Bogdanov, A. A. (2002). Mouse chromocenters contain associated telomeric DNA and telomerase activity. *Doklady Biological Sciences: Proceedings of the Academy of Sciences of the USSR, Biological Sciences Sections, 383*, 171–174.

Enukashvily, N., Donev, R., Sheer, D., & Podgornaya, O. (2005). Satellite DNA binding and cellular localisation of RNA helicase P68. *Journal of Cell Science, 118*, 611–622.

Enukashvily, N. I., Donev, R., Waisertreiger, I. S.-R., & Podgornaya, O. I. (2007). Human chromosome 1 satellite 3 DNA is decondensed, demethylated and transcribed in senescent cells and in A431 epithelial carcinoma cells. *Cytogenetic and Genome Research, 118*(1), 42–54.

Enukashvily, N. I., Lobov, I. B., Kukalev, A., & Podgornaya, O. I. (2000). A nuclear matrix protein related to intermediate filaments proteins is a member of the complex binding alphoid DNA in vitro. *Cell Biology International, 24*(7), 483–492.

Enukashvily, N. I., Malashicheva, A. B., & Waisertreiger, I. S.-R. (2009). Satellite DNA spatial localization and transcriptional activity in mouse embryonic E-14 and IOUD2 stem cells. *Cytogenetic and Genome Research, 124*(3–4), 277–287.

Fitzgerald, D. J., Dryden, G. L., Bronson, E. C., Williams, J. S., & Anderson, J. N. (1994). Conserved patterns of bending in satellite and nucleosome positioning DNA. *The Journal of Biological Chemistry, 269*(33), 21303–21314.

Garagna, S., Zuccotti, M., Capanna, E., & Redi, C. A. (2002). High-resolution organization of mouse telomeric and pericentromeric DNA. *Cytogenetic and Genome Research, 96*(1–4), 125–129.

Giordano, J., Ge, Y., Gelfand, Y., Abrusan, G., Benson, G., & Warburton, P. E. (2007). Evolutionary history of mammalian transposons determined by genomewide defragmentation. *PLoS computational biology, 3*, e137.

Gosden, J. R., Mitchell, A. R., Buckland, R. A., Clayton, R. P., & Evans, H. J. (1975). The location of four human satellite DNAs on human chromosomes. *Experimental Cell Research*, *92*(1), 148–158.

Grewal, S. I. S., & Elgin, S. C. R. (2007). Transcription and RNA interference in the formation of heterochromatin. *Nature*, *447*(7143), 399–406.

Guenatri, M., Bailly, D., Maison, C., & Almouzni, G. (2004). Mouse centric and pericentric satellite repeats form distinct functional heterochromatin. *The Journal of Cell Biology*, *166*(4), 493–505.

Haaf, T., & Ward, D. C. (1996). Inhibition of RNA polymerase II transcription causes chromatin decondensation, loss of nucleolar structure, and dispersion of chromosomal domains. *Experimental Cell Research*, *224*(1), 163–173.

Hörz, W., & Altenburger, W. (1981). Sequence specific cleavage of DNA by micrococcal nuclease. *Nucleic Acids Research*, *9*(12), 2643–2658.

Hörz, W., & Zachau, H. G. (1977). Characterization of distinct segments in mouse satellite DNA by restriction nucleases. *European Journal of Biochemistry*, *73*(2), 383–392.

Jabs, E. W., Wolf, S. F., & Migeon, B. R. (1984). Characterization of a cloned DNA sequence that is present at centromeres of all human autosomes and the X chromosome and shows polymorphic variation. *Proceedings of the National Academy of Sciences of the United States of America*, *81*(15), 4884–4888.

Joseph, A., Mitchell, A. R., & Miller, O. J. (1989). The organization of the mouse satellite DNA at centromeres. *Experimental Cell Research*, *183*(2), 494–500.

Kalitsis, P., Griffiths, B., & Choo, K. H. A. (2006). Mouse telocentric sequences reveal a high rate of homogenization and possible role in Robertsonian translocation. *Proceedings of the National Academy of Sciences of the United States of America*, *103*(23), 8786–8791.

Kipling, D. (1995). Telomerase: Immortality enzyme or oncogene? *Nature Genetics*, *9*(2), 104–106.

Kipling, D., Ackford, H. E., Taylor, B. A., & Cooke, H. J. (1991). Mouse minor satellite DNA genetically maps to the centromere and is physically linked to the proximal telomere. *Genomics*, *11*(2), 235–241.

Kireeva, N., Lakonishok, M., Kireev, I., Hirano, T., & Belmont, A. S. (2004). Visualization of early chromosome condensation: A hierarchical folding, axial glue model of chromosome structure. *The Journal of Cell Biology*, *166*(6), 775–785.

Kit, S. (1961). Equilibrium sedimentation in density gradients of DNA preparations from animal tissues. *Journal of Molecular Biology*, *3*, 711–716.

Kit, S. (1962). Species differences in animal deoxyribonucleic acid as revealed by equilibrium sedimentation in density gradients. *Nature*, *193*, 274–275.

Kobliakova, I., Zatsepina, O., Stefanova, V., Polyakov, V., & Kireev, I. (2005). The topology of early- and late-replicating chromatin in differentially decondensed chromosomes. *Chromosome Research*, *13*(2), 169–181.

Komissarov, A. S., Gavrilova, E. V., Demin, S. J., Ishov, A. M., & Podgornaya, O. I. (2011). Tandemly repeated DNA families in the mouse genome. *BMC Genomics*, *12*, 531.

Komissarov, A. S., Kuznetsova, I. S., & Podgornaia, O. I. (2010). Mouse centromeric tandem repeats in silico and in situ. *Genetika*, *46*(9), 1217–1221.

Kukalev, A. S., Enukashvili, N. I., & Podgornaia, O. I. (2005). Multifunctional nuclear protein NAP57 specifically interacts with dead RNA-helicase p68. *Tsitologiia*, *47*(6), 533–539.

Kukalev, A. S., Lobov, I. B., Percipalle, P., & Podgornaya, O. I. (2009). SAF-A/hnRNP-U localization in interphase and metaphase. *Cytogenetic and Genome Research*, *124*(3–4), 288–297.

Kukalev, A., Nord, Y., Palmberg, C., Bergman, T., & Percipalle, P. (2005). Actin and hnRNP U cooperate for productive transcription by RNA polymerase II. *Nature Structural & Molecular Biology*, 238–244.

Kuznetsova, I. S., Voronin, A. P., & Podgornaya, O. I. (2006). Telomere and TRF2/MTBP localization in respect to satellite DNA during the cell cycle of mouse cell line L929. *Rejuvenation Research, 9*(3), 391–401.

Lee, C., Wevrick, R., Fisher, R. B., Ferguson-Smith, M. A., & Lin, C. C. (1997). Human centromeric DNAs. *Human Genetics, 100*(3–4), 291–304.

Lengyel, P., Speyer, J. F., & Ochoa, S. (1961). Synthetic polynucleotides and the amino acid code. *Proceedings of the National Academy of Sciences of the United States of America, 47,* 1936–1942.

Lobov, I. B., Tsutsui, K., Mitchell, A. R., & Podgornaya, O. I. (2000). Specific interaction of mouse major satellite with MAR-binding protein SAF-A. *European Journal of Cell Biology, 79,* 839–849.

Lobov, I. B., Tsutsui, K., Mitchell, A. R., & Podgornaya, O. I. (2001). Specificity of SAF-A and lamin B binding in vitro correlates with the satellite DNA bending state. *Journal of Cellular Biochemistry, 83*(2), 218–229.

López-Flores, I., & Garrido-Ramos, M. A. (2012). The repetitive DNA content of eukaryotic genomes. *Genome Dynamics, 7,* 1–28.

Lu, J., & Gilbert, D. M. (2007). Proliferation-dependent and cell cycle regulated transcription of mouse pericentric heterochromatin. *The Journal of Cell Biology, 179*(3), 411–421.

Mahtani, M. M., & Willard, H. F. (1990). Pulsed-field gel analysis of alpha-satellite DNA at the human X chromosome centromere: High-frequency polymorphisms and array size estimate. *Genomics, 7*(4), 607–613.

Maison, C., Bailly, D., Peters, A. H. F. M., Quivy, J.-P., Roche, D., Taddei, A., et al. (2002). Higher-order structure in pericentric heterochromatin involves a distinct pattern of histone modification and an RNA component. *Nature Genetics, 30*(3), 329–334.

Mamaeva, S. (2002). *Atlas of chromosomes of human and animal cell lines.* Moscow: Scentific World.

Manuelidis, L. (1976). Repeating restriction fragments of human DNA. *Nucleic Acids Research, 3*(11), 3063–3076.

Manuelidis, L. (1978a). Chromosomal localization of complex and simple repeated human DNAs. *Chromosoma, 66*(1), 23–32.

Manuelidis, L. (1978b). Complex and simple sequences in human repeated DNAs. *Chromosoma, 66*(1), 1–21.

Manuelidis, L., & Borden, J. (1988). Reproducible compartmentalization of individual chromosome domains in human CNS cells revealed by in situ hybridization and three-dimensional reconstruction. *Chromosoma, 96*(6), 397–410.

Markova, D., Donev, R., Patriotis, C., & Djondjurov, L. (1994). Interphase chromosomes of Friend-S cells are attached to the matrix structures through the centromeric/telomeric regions. *DNA and Cell Biology, 13*(9), 941–951.

Martienssen, R. A. (2003). Maintenance of heterochromatin by RNA interference of tandem repeats. *Nature Genetics, 35*(3), 213–214.

Martínez-Balbás, A., Rodríguez-Campos, A., García-Ramírez, M., Sainz, J., Carrera, P., Aymamí, J., et al. (1990). Satellite DNAs contain sequences that induced curvature. *Biochemistry, 29*(9), 2342–2348.

Mayer, C., Leese, F., & Tollrian, R. (2010). Genome-wide analysis of tandem repeats in Daphnia pulex—A comparative approach. *BMC Genomics, 11,* 277.

Moyzis, R. K., Albright, K. L., Bartholdi, M. F., Cram, L. S., Deaven, L. L., Hildebrand, C. E., et al. (1987). Human chromosome-specific repetitive DNA sequences: Novel markers for genetic analysis. *Chromosoma, 95*(6), 375–386.

Muchardt, C., Guilleme, M., Seeler, J.-S., Trouche, D., Dejean, A., & Yaniv, M. (2002). Coordinated methyl and RNA binding is required for heterochromatin localization of mammalian HP1alpha. *EMBO Reports, 3*(10), 975–981.

Narayanswami, S., Doggett, N. A., Clark, L. M., Hildebrand, C. E., Weier, H. U., & Hamkalo, B. A. (1992). Cytological and molecular characterization of centromeres in Mus domesticus and Mus spretus. *Mammalian Genome: Official Journal of the International Mammalian Genome Society*, 2(3), 186–194.

Paar, V., Pavin, N., Rosandic, M., Gluncic, M., Basar, I., Pezer, R., et al. (2005). ColorHOR—Novel graphical algorithm for fast scan of alpha satellite higher-order repeats and HOR annotation for GenBank sequence of human genome. *Bioinformatics (Oxford, England)*, 21(7), 846–852.

Palomeque, T., & Lorite, P. (2008). Satellite DNA in insects: A review. *Heredity*, 100(6), 564–573.

Pardue, M. L., & Gall, J. G. (1970). Chromosomal localization of mouse satellite DNA. *Science*, 168, 1356–1358.

Pertile, M. D., Graham, A. N., Choo, K. H. A., & Kalitsis, P. (2009). Rapid evolution of mouse Y centromere repeat DNA belies recent sequence stability. *Genome Research*, 19(12), 2202–2213.

Pezer, Z., Brajković, J., Feliciello, I., & Ugarković, D. (2012). Satellite DNA-mediated effects on genome regulation. *Genome Dynamics*, 7, 153–169.

Phair, R. D., & Misteli, T. (2000). High mobility of proteins in the mammalian cell nucleus. *Nature*, 404(6778), 604–609.

Phair, R. D., Scaffidi, P., Elbi, C., Vecerová, J., Dey, A., Ozato, K., et al. (2004). Global nature of dynamic protein-chromatin interactions in vivo: Three-dimensional genome scanning and dynamic interaction networks of chromatin proteins. *Molecular and Cellular Biology*, 24(14), 6393–6402.

Pietras, D. F., Bennett, K. L., Siracusa, L. D., Woodworth-Gutai, M., Chapman, V. M., Gross, K. W., et al. (1983). Construction of a small Mus musculus repetitive DNA library: Identification of a new satellite sequence in Mus musculus. *Nucleic Acids Research*, 11(20), 6965–6983.

Podgornaia, O. I., Ostromyshenskiĭ, D. I., Kuznetsova, I. S., Matveev, I. V., & Komissarov, A. S. (2009). Heterochromatin and centromere structure paradox. *Tsitologiia*, 51(3), 204–211.

Podgornaya, O. I., Bugaeva, E. A., Voronin, A. P., Gilson, E., & Mitchell, A. R. (2000). Nuclear envelope associated protein that binds telomeric DNAs. *Molecular Reproduction and Development*, 57, 16–25.

Podgornaya, O., Dey, R., Lobov, I., & Enukashvili, N. (2000). Human satellite 3 (HS3) binding protein from the nuclear matrix: Isolation and binding properties. *Biochimica et Biophysica Acta*, 1497(2), 204–214.

Podgornaya, O. I., Voronin, A. P., Enukashvily, N. I., Matveev, I. V., & Lobov, I. B. (2003). Structure-specific DNA-binding proteins as the foundation for three-dimensional chromatin organization. *International Review of Cytology*, 224, 227–296.

Probst, A. V., Okamoto, I., Casanova, M., El Marjou, F., Le Baccon, P., & Almouzni, G. (2010). A strand-specific burst in transcription of pericentric satellites is required for chromocenter formation and early mouse development. *Developmental Cell*, 19(4), 625–638.

Prosser, J., Frommer, M., Paul, C., & Vincent, P. C. (1986). Sequence relationships of three human satellite DNAs. *Journal of Molecular Biology*, 187(2), 145–155.

Richard, G.-F., Kerrest, A., & Dujon, B. (2008). Comparative genomics and molecular dynamics of DNA repeats in eukaryotes. *Microbiology and Molecular Biology Reviews*, 72(4), 686–727.

Rudd, M. K., & Willard, H. F. (2004). Analysis of the centromeric regions of the human genome assembly. *Trends in Genetics*, 20(11), 529–533.

Schueler, M. G., Higgins, A. W., Rudd, M. K., Gustashaw, K., & Willard, H. F. (2001). Genomic and genetic definition of a functional human centromere. *Science (New York, N.Y.)*, 294(5540), 109–115.

Smit, A. F. (1993). Identification of a new, abundant superfamily of mammalian LTR-transposons. *Nucleic Acids Research*, *21*(8), 1863–1872.

Solovjeva, L. V., Demin, S. J., Pleskach, N. M., Kuznetsova, M. O., & Svetlova, M. P. (2012). Characterization of telomeric repeats in metaphase chromosomes and interphase nuclei of Syrian Hamster Fibroblasts. *Molecular Cytogenetics*, *5*(1), 37.

Stocking, C., & Kozak, C. A. (2008). Murine endogenous retroviruses. *Cellular and Molecular Life Sciences*, *65*(21), 3383–3398.

Trifonov, E. N., & Sussman, J. L. (1980). The pitch of chromatin DNA is reflected in its nucleotide sequence. *Proceedings of the National Academy of Sciences of the United States of America*, *77*(7), 3816–3820.

Ushiki, T., & Hoshi, O. (2008). Atomic force microscopy for imaging human metaphase chromosomes. *Chromosome Research*, *16*(3), 383–396.

Valgardsdottir, R., Chiodi, I., Giordano, M., Rossi, A., Bazzini, S., Ghigna, C., et al. (2008). Transcription of Satellite III non-coding RNAs is a general stress response in human cells. *Nucleic Acids Research*, *36*(2), 423–434.

Vissel, B., & Choo, K. H. (1989). Mouse major (gamma) satellite DNA is highly conserved and organized into extremely long tandem arrays: Implications for recombination between nonhomologous chromosomes. *Genomics*, *5*(3), 407–414.

Vogt, P. (1990). Potential genetic functions of tandem repeated DNA sequence blocks in the human genome are based on a highly conserved "chromatin folding code" *Human Genetics*, *84*(4), 301–336.

Voronin, A. P., Lobov, I. B., Bugaeva, E. A., Parfenov, V. N., & Podgornaia, O. I. (1999a). Telomere-binding protein of mouse nuclear matrix. II. Localization. *Molekuliarnaia Biologiia*, *33*(4), 665–672.

Voronin, A. P., Lobov, I. B., Bugaeva, E. A., Parfenov, V. N., & Podgornaia, O. I. (1999b). Telomere-binding protein of mouse nuclear matrix. I. Characteristics. *Molekuliarnaia Biologiia*, *33*(4), 657–664.

Warburton, P. E., Hasson, D., Guillem, F., Lescale, C., Jin, X., & Abrusan, G. (2008). Analysis of the largest tandemly repeated DNA families in the human genome. *BMC Genomics*, *9*, 533.

Waterston, R. H., Lindblad-Toh, K., Birney, E., Rogers, J., Abril, J. F., et al. (2002). Initial sequencing and comparative analysis of the mouse genome. *Nature*, *420*(6915), 520–562.

White, S. A., & Allshire, R. C. (2008). RNAi-mediated chromatin silencing in fission yeast. *Current Topics in Microbiology and Immunology*, *320*, 157–183.

Wong, A. K., & Rattner, J. B. (1988). Sequence organization and cytological localization of the minor satellite of mouse. *Nucleic Acids Research*, *16*(24), 11645–11661.

Yunis, J. J., & Yasmineh, W. G. (1971). Heterochromatin, satellite DNA, and cell function. Structural DNA of eucaryotes may support and protect genes and aid in speciation. *Science*, *174*(4015), 1200–1209.

CHAPTER TWO

Mammalian Satellite DNA: A Speaking Dumb

Natella I. Enukashvily[1], Nikita V. Ponomartsev
Institute of Cytology RAS, St. Petersburg, Russia
[1]Corresponding author. e-mail address: natella@mail.cytspb.rssi.ru

Contents

Abstract

The tandemly organized highly repetitive satellite DNA is the main DNA component of centromeric/pericentromeric constitutive heterochromatin. For almost a century, it was considered as "junk DNA," only a small portion of which is used for kinetochore formation. The current review summarizes recent data about satellite DNA transcription. The possible functions of the transcripts are discussed.

1. INTRODUCTION

In 1928, Heitz suggested the terms euchromatin and heterochromatin (HC) for differences detectable by suitable chromosomal stains (Heitz, 1928). He stained cells from several species of moss with carmine acetic acid and observed a type of chromatin in the nucleus that remained condensed throughout the cell cycle. Heitz described that portion of the nuclear chromatin as heterochromatin, which maintained a condensed state

Advances in Protein Chemistry and Structural Biology, Volume 90
ISSN 1876-1623
http://dx.doi.org/10.1016/B978-0-12-410523-2.00002-X

(i.e., appeared darkly stained) throughout the cell interphase, while the remainder of the nuclear chromatin was extending to what he termed the euchromatin state. Cooper (1959) was able to summarize the data from Drosophila and suggested that heterochromatin and euchromatin differed in their biophysical conformations and in metabolic expression of their genes, but not in their basic structure of DNA arranged within chromosomes. Now, the concept of a eukaryotic genome consisting of two types of differently packed chromatin is widely accepted and it is included in school textbooks in biology. Although the term heterochromatin was originally defined cytologically as regions of mitotic chromosomes that remain condensed in the interphase, it is now more loosely applied to include regions of chromosomes that show characteristic properties such as, for example, gene repression and silencing (reviewed by Craig, 2005; Lohe & Hilliker, 1995).

It is commonly accepted that there are two types of heterochromatin—constitutive and facultative. The constitutive HC is thought to be condensed throughout the entire cell cycle unlike the facultative HC which is developmentally regulated. In mitotic chromosomes, constitutive heterochromatic regions are positioned most often in centromeric and pericentromeric regions as well as near telomeres. Chromosomes 1, 9, 16, and Y contain large blocks of heterochromatin. The facultative HC can be formed on different chromosomes regions. In mammalian females, one X-chromosome (either maternally or paternally derived) is randomly inactivated in early embryonic cells, with fixed inactivation in all descendant cells (Lyon, 1961). Facultative HC can be formed in the promoter regions of non-transcribed genes (Rand & Cedar, 2003). Some of the neocentromeres (regions of euchromatic DNA that acquired properties of centromeres) known to date can undergo heterochromatinization though they are formed within euchromatic regions (Amor & Choo, 2002; Saffery et al., 2003).

2. HETEROCHROMATIN CRITERIA

The following criteria are often used to classify a chromosome region as a region belonging to constitutive HC:

1. The main morphological criterion is that the constitutive HC remains condensed throughout the cell cycle (Heitz, 1928).
2. Constitutive HC lacks of genes. Moreover, constitutive HC can induce variegated expression of euchromatic genes when two types of chromatin were abnormally juxtaposed by chromosome rearrangements. This phenomenon, called position effect variegation (PEV), was discovered

by Muller (1930). Further studies revealed that dozens of euchromatic genes are inactivated when placed near or in heterochromatin (Baker, 1968). The generality of PEV showed that heterochromatin not only lacks gene activity but could also routinely cause gene inactivation.

3. Constitutive heterochromatin largely comprises highly repetitive satellite sequences and middle-repetitive transposable elements (TEs; reviewed in Choo, 1997). These classes of repetitive DNA had no known biological function and were therefore sometimes referred to as "junk DNA." Ohno (1972) is credited with originating the term. However, in fact, his paper was focused mainly on the fossilized genes. But as the term caught on in the 1980s, its meaning was extended to "all noncoding sequences, the vast stretches of DNA that are not genes and do not produce proteins."

4. Constitutive HC is strongly stained with intercalating dyes such as Hoechst or DAPI (Schnedl, Mikelsaar, Breitenbach, & Dann, 1977).

5. The regions of HC are enriched in Heterochromatin protein 1 (HP1) or its isophorms (in mammalians: HP1α, HP1β, HP1γ) (Craig, 2005).

6. Constitutive HC DNA is late replicating in the S-phase (reviewed in Gilbert, 2002). Experiments conducted in the early 1960s showed that genetically inactive heterochromatin, contained in Giemsa dark chromosome bands or "G bands," replicates late during S-phase, whereas most transcription takes place in Giemsa light or "R bands," which replicate early during S-phase. Results from modern molecular approaches have been consistent with this conclusion: out of the few dozen genes examined, nearly all transcriptionally active genes replicate early in S-phase, and more than half of the developmentally regulated genes replicate late when they are not expressed.

7. Some of the epigenetic modifications are a hallmark of HC. 5-methylcytosine (5mC) is one of them. Postreplicative methylation of cytosine is carried out by a diverse group of DNA methyltransferases. Some of these are capable of de novo methylation, whereas others primarily maintain the pattern of methylation by acting preferentially on hemimethylated sites after replication. CpG is the most common site of methylation, but CpNpG sites and asymmetric sites are also used (Fazzari & Greally, 2004; Grewal and Elgin, 2007). Most of the constitutive HC DNAs are methylated in 5mC in CpG islands. In human normal somatic tissues, most (70–90%) of the CpG islands are methylated (Wilson, Power, & Molloy, 2007). Proteins binding to methylated DNA of CpG islands (Met-CpG-binding proteins) are usually also methylated especially in pericentromeric HC (Maison et al., 2002; Peters et al., 2003).

The methylation of CpG islands is a modification of DNA itself. However, different histone modifications are also involved in heterochromatin formation. Histones may undergo post-translational modifications of their N-terminal ends which may affect the genetic activity of the chromatin. The hypoacetylation of histone N-terminal tails, principally on the lysines, is associated with an inactive chromatin. In contrast, hyperacetylated histones characterize the active chromatin. Acetylation/deacetylation of histones is a mechanism that is absolutely essential for the control of gene expression. Numerous transcription factors have been shown to have either an activity histone acetyl transferase or histone deacetylases (Grewal and Elgin, 2007). Heterochromatin is also characterized by methylation of histone H3 at lysine 9 in higher eukaryotes, but not in some single-celled eukaryotes such as Saccharomyces. The studies by Peters et al. (2003) and Rice et al. (2003) showed that H3 K9 trimethylation and H3 K27 monomethylation are enriched in pericentromeric heterochromatin. Histone H3 methylated at lysine 9 (H3-mK9) is bound by Heterochromatin protein 1 (HP1). Histone methylation at lysine residues is central to epigenetic regulation of gene expression and is carried out by histone methyltransferases (Craig, 2005; Dillon, 2004; Grewal and Elgin, 2007). These three major biochemical marks—histone hypoacetylation, H3-K9 methylation, and DNA methylation—are considered now as the main biochemical markers of heterochromatin.

Another very important epigenetic modification that is a hallmark of constitutive HC is the recruitment of CENP-A nucleosomes to existing centromeres. The recruitment is independent of the underlying DNA sequence. Early experiments using antisera from patients with CREST syndrome identified CENP-A as a 17-kDa protein (reviewed in Choo, 1997). The protein shares a similar organization with the histone H3 (Sullivan, Hechenberger, & Masri, 1994). Heterochromatin is frequently found near CENP-A chromatin, which is the key determinant of kinetochore assembly. In fission yeast, *de novo* recruitment of the CENP-A to the centromere is believed to be controlled by "centromeric" heterochromatin surrounding the centromere, and by an RNAi mechanism. Once assembled, CENP-A chromatin is propagated by epigenetic means in the absence of heterochromatin (Folco, Pidoux, Urano, & Allshire, 2008).

None of the criteria listed above can be used as a stand-alone criterion. For example, DAPI (or other intercalating dyes) is not a precise marker of heterochromatic foci in the mouse nucleus (Guenatri, Bailly, Maison, & Almouzni, 2004). The HP1-protein is bound to pericentromeric heterochromatin but not to the inactive X-chromosome where it is absent (Kellum et al., 2003). Constitutive HC can interact with PcG (Polycomb Group) proteins (Maison et al., 2002; Peters et al., 2003). The decodensation of constitutive HC is also demonstrated (in details see Section 4.1). Central domains of human α-satellite DNA are partially decondensed in prometaphase. The maximum of decondensation is observed in metaphase when mitotic spindle microtubules are attached to centromeres. Similar decondensation during mitosis was demonstrated in other species as well (Shelby, Hahn, & Sullivan, 1996; Skibbens, Skeen, & Salmon, 1993). In mouse cells, decondensed threads of late-replicating satellite DNA emanating from condensed centromeres and connecting adjacent chromosomes were observed by several authors (referenced in Kuznetsova et al., 2007). The decondensation of heterochromatin was also observed in some malignant or aging cell cultures (De Cecco et al., 2013; Enukashvily, Donev, Waisertreiger, & Podgornaya, 2007). The transcription of satellite DNA underlying constitutive HC is now a well-accepted fact (Metz, Soret, Vourc'h, Tazi, & Jolly, 2004; Valgardsdottir et al., 2005; Vourc'h & Biamonti, 2011 and the current review). Moreover, partial or full demethylation of constitutive HC DNA is established. The hypomethylation of pericetromeric satellite 2 and/or α was demonstrated in cells from patients with ICF syndrome and in cells obtained from different tumors including Wilms tumors, breast and ovary carcinomas (Enukashvily et al., 2007; Wilson et al., 2007). Thus, even the "classical" fundamental criteria of heterochromatin (condensation, transcriptional silence, hypermethylation) are not absolute and all of the criteria above should be taken into account referring a region as belonging to constitutive HC.

3. THE CONSTITUTIVE HC DNA

3.1. Satellite DNA

Human centromeric HC is formed on the basis of chromosome-specific alphoid (or α-satellite) DNA. Some other types of satellite DNA such as γ-satellites and Sn5-tandem repeats were also found in centromeres regions (Choo, 1997; Lin, Sasi, Lee, Fan, & Court, 1993). Vast arrays of centromeric alphoid DNA consist of tandemly organized repeats with a monomer length of 171 bp. It was assumed that alphoid DNA is the only type of DNA that resides in centromere, underlies kinetochore, and is sufficient for assembling kinetochore *de novo* and maintaining human artificial chromosome stability (Alazami, Mejia, & Monaco, 2004; Basu & Willard, 2005; Henning et al., 1999; Ikeno et al., 1998; Okada et al., 2007). However these artificial chromosomes are not stable in mitosis, their size is always larger than the original construct and the sequence of acquired DNAs is unknown (Podgornaia, Ostromyshenskiĭ, Kuznetsova, Matveev, & Komissarov, 2009). It is obvious now that some other yet unknown sequences are necessary for centromere functioning and kinetochore assembly.

The centromeres of other mammalian are also built of satellite DNA. In mice, the DNA of centromeric regions of all chromosomes (except Y-chromosome) is comprised of minor satellite DNA (Mitchell, 1996). The blocks of tandemly organized minor satellite monomers are intermingled with blocks of another satellite (Kuznetsova, Podgornaya, & Ferguson-Smith, 2006; Kuznetsova, Prusov, Enukashvily, & Podgornaya, 2005). The presence of other satellites within mouse centromeres is reported (Komissarov, Gavrilova, Demin, Ishov, & Podgornaya, 2011).

The regions juxtaposing centromeres–pericentromeric regions are also involved in centromere functioning (Bernard & Allshire, 2002; Durand-Dubief & Ekwall, 2008). In human, a host of repetitive DNA and α-satellite DNA have been localized to pericentric regions of human chromosomes. The classical satellites, satellites 1–3, are among the first tandemly repeated DNA characterized in the human genome. These satellite sequences are made of short repeats that are distinguishable from the rest of the genomic DNA because they form satellite bands on caesium chloride density gradient (Choo, 1997; Lee, Wevrick, Fisher, Ferguson-Smith, & Lin, 1997; Vissel, Nagy, & Choo, 1992). In mouse species, major satellite is known to be associated close to centromeres of mouse acrocentric chromosomes (Vissel and Choo, 1989). Another satellite—MS4—was also found within major satellite arrays (Kuznetsova et al., 2005, 2006).

3.2. Non-satellite centromeric and pericentromeric DNA

Tandemly organized highly repetitive satellite DNA is not the only type of DNA residing within centromeric and pericentromeric regions of higher eukaryotes. Some regulatory sequences, TEs, genes or gene fragments were found within centromeric and especially pericentromeric HC DNA. Some of them can be regulatory elements of satellite DNA transcription (Carone et al., 2009; Kuznetsova et al., 2006; O'Neill and Carone, 2009). Transcriptionally active genes were found within centromeric regions of *Drosophila melanogaster*. Moreover, these genes must be surrounded by heterochromatin to function properly; transferring them into euchromatin will impair their expression (Howe, Dimitri, Berloco, & Wakimoto, 1995). Smith, Shu, Mungall, and Karpen (2007) demonstrated that the heterochromatin contained a minimum of 230–254 protein-coding genes, which are conserved in other Drosophilids and more diverged species, as well as 32 pseudogenes and 13 noncoding RNAs. Improved methods revealed that more than 77% of this heterochromatin sequence, including introns and intergenic regions, was composed of fragmented and nested TEs and other repeated DNAs. In rice, the chromosome 8 centromere also contains transcriptionally active genes that are adjacent to kinetochore satellite DNA (Plohl, Luchetti, Mestrović, & Mantovani, 2008; Yan & Jiang, 2007). In S. pombe, tRNA genes were mapped at the boundary between centromeric and pericentromeric heterochromatin (Partridge, Borgstrom, & Allshire, 2000). Some genes (such as *BAGE* on chromosomes 13 and 21) were mapped to juxtacentromeric regions in human. Ruault et al. (2002) published the data confirming that the *BAGE*-related sequences are located less than 1 Mb away from the centromere: two in 21p and one in 21q. The authors showed that *BAGE* is a gene family composed of expressed genes that map to the juxtacentromeric regions of chromosomes 13 and 21, and of unexpressed gene fragments that are scattered in the juxtacentromeric regions of several chromosomes. Tracts of gene-related sequences have been identified proximal to satellite 2 in pericentromere of chromosomes 22 and at the terminus of alphoid DNA array on chromosome 16 (Dunham et al., 1999; Horvath et al., 2000) as well as in chromosome 21 pericentromere (Brun, Ruault, Ventura, Roizès, & De Sario, 2003). Three genes were mapped within the boundary region between satellite-rich centromere and chromosome-specific euchromatin on human chromosome 10 (Guy et al., 2000). The identification of a partial duplication of *GABRA5*, a gene within the imprinted 15q11–q13 region, was reported. The duplicated locus maps to the pericentromeric region of 15q and might influence transcription of adjacent satellite DNA (Ritchie, Mattei, & Lalande, 1998). The transcriptional promoter of satellite 2 from the eastern newt, Notophthalmus viridescens is similar to the sequences of the octamer and the proximal sequence element of vertebrate small nuclear RNA (snRNA) genes. The satellite 2 promoter initiated transcription of the U1b2 snRNA gene as efficiently as the native U1b2 promoter (Coats, Zhang, & Epstein, 1994).

Other sequences beside heterochromatic genes and their fragments that can be found within pericentromeres belong to TEs family. In mouse, two families that have structural similarity to TEs were mapped to centromere and/or pericentromere (Komissarov et al., 2011). First family, TR–MTA, is formed by MTA fragments: MTa, MaLR-LTR, and Mammalian apparent LTR-retrotransposons in Repbase, whereas second family, L1-related family, is formed by part of the ORF2 and 3′LTR. MTA transposons have structural similarities to endogenous retroviruses, namely ERV3, and are related to THE1 in humans. Preliminary analysis has not yet revealed significant similarities of the putative product of MTA ORF to any protein present in the databanks. The existence of both L1 and MTA arrays within centromeric and pericentromeric DNA *is predicted in silico* (Chapter 1).

For the past half century, no function was attributed to all this bulk of DNA except a single one—kinetochore formation and sister-chromatin cohesion. As only a small portion of centromeric DNA was shown to be necessary to accomplish this function, this entire DNA (comprising 5–15% of genome) was termed as "junk DNA" along with all other noncoding DNA (98.7% of human genome). However, in the past two decades, this point of view was abandoned as new data were obtaining confirming the importance of noncoding DNA. Now the term "genome dark matter" (Pennisi, 2010) is used much more often than almost abandoned "junk DNA."

4. TRANSCRIPTION OF HC DNA

4.1. Early data on constitutive HC DNA decondensation and transcription

Transcriptional silence and high degree of condensation were believed to be fundamental attributes of constitutive HC for several decades. Creating the environment for kinetochore formation was assumed to be the only function of satellite DNA. However, at the end of 1960s, the first data appeared that gave a hint about possible satellite DNA transcription. Harel, Hanania, Tapiero, and Harel (1968) detected a rapidly labeled RNA in liver, kidney, and L–cells, capable of hybridizing with the deoxyadenylate-rich strand of satellite DNA. Cohen, Rode, and Helleiner (1972) and Cohen, Huh, and Helleiner (1973) gave evidence that mouse L-cells contain RNA capable of hybridizing with mouse satellite DNA on nitrocellulose membranes and of competing for hybridizing sites on satellite DNA with transcript of satellite DNA prepared *in vitro*. Seidman and Cole (1977) showed that pulse-labeled RNA was binding satellite DNA as well. However, they did not believe themselves and concluded that equal amounts of satellite DNA in all fractions of mouse L-cell chromatin indicated that the method of sheared chromatin centrifugation in sucrose gradient did not fractionate on the basis of the *in vivo* transcriptional activity. Nevertheless, transcription of satellite DNA was demonstrated *in situ* in lampbrush chromosomes of *Triturus cristatus* some years later (Macgregor, 1979; Varley, Macgregor,

Nardi, Andrews, & Erba, 1980). Transcription of lampbrush chromosomes pericentromeres was further confirmed in experiments on different amphibians (Diaz, Barsacchi-Pilone, Mahon, & Gall, 1981; Jamrich, Warrior, Steele, & Gall, 1983) and birds (Krasikova, Vasilevskaia, & Gaginskaia, 2010). Today, the transcription of centromeric and/or pericentromeric DNA is confirmed in many species. Transcripts were found in cells of many Hymenoptera species (Rouleux-Bonnin, Renault, Bigot, & Periquet, 1996), *Diprion pini*, *Bombus terrestris* (Rouleux-Bonnin, Bigot, & Bigot, 2004), and *Aphaenogaster subterra* (Lorite et al., 2002). Satellite DNA transcripts were found in Diptera species such as *Drosophila melanogaster* (Bonaccorsi, Gatti, Pisano, & Lohe, 1990) and *Drosophila hydei* (Trapitz, Wlaschek, & Bünemann, 1988). The data about transcription of satellite DNA became numerous to date. These findings suggest that the phenomenon of satellite DNA transcription is common (Lee, Neumann, Macas, & Jiang, 2006; Lehnertz et al., 2003; Lorite et al., 2002; Martienssen, 2003; Pathak et al., 2006; Rizzi et al., 2004; Rudert, Bronner, Garnier, & Dollé, 1995; Valgardsdottir et al., 2008; Varadaraj & Skinner, 1994).

Transcription of a chromatin region is often accompanied by its decondensation. Indeed the phenomenon of constitutive HC regions decondensation and reorganization was described in different cell models both on mitotic and interphase chromosomes. The decondensation of the chromosome 1 pericentromeric region in mitosis was described by Donahue, Bias, Renwick, and McKusick (1968) as a case of marker chromosome occurring in lymphocytes of family V11DE. Pérez et al. (1991) observed spontaneous decondensation of chromosomes 1, 9, 16, and Y centromeric HC in 46% of tested chorionic villus samples. In interphase, a remarkable decondensation of chromosome 1 pericentromere was observed in A431 tumor cell line and in MRC5 senescent fibroblasts (Enukashvily et al., 2007). The authors assumed that the interphase decondensation made the binding sites for transcription factors present in HS3-1 accessible as the transcription occurred only in cell lines with decondensed pericentromeric HC but not in cell lines with condensed pericentromeric HC. The decondensation of pericentromeres was observed also in other tumor and senescent cell cultures (De Cecco et al., 2013; Narayan et al., 1997; Suzuki, Fujii, & Ayusawa, 2002). Human alphoid DNA is shown to decondense in mitosis before the mitotic spindle microtubules attachment. Similar decondensation is shown in other species (Shelby et al., 1996; Skibbens et al., 1993). Centromeric DNA of Indian muntjack starts to decondense at the S-phase before replication (He and Brinkley, 1996). In mouse cells, threads of late-replicating satellite DNA emanate from compact centromeres (Kuznetsova et al., 2007). In insects, satellite DNA starts to decondense under stress conditions (Campos, Rodrigues, Wada, & Mello, 2002; Dantas & Mello, 1992). Similar data were obtained in rice. In *Arabidopsis thaliana* and *Nicotiana tabacum*, protoplasts dedifferentiation is accompanied with heterochromatin decondensation and chromocenters reorganization (Avivi et al., 2004; Tessadori, Chupeau, et al., 2007; Zhao et al., 2001). In Arabidopsis, similar processes take place during floral transition (Tessadori, Schulkes, van Driel, & Fransz, 2007).

4.2. Transcription of constitutive HC DNA in mammals

4.2.1 Heterochromatic genes

Some genes are specifically adapted to be expressed exclusively in a heterochromatic context (heterochromatic genes). The study of heterochromatic genes began in the early days of Drosophila genetics (Sturtevant, 1965). In 1916, Calvin Bridges suggested that the Y-chromosome must carry genes necessary for male fertility, although, as realized later, this chromosome was considered entirely heterochromatic by cytological definition. In addition, several mutations, including the *bobbed*, *light*, and *rolled* mutations, mapped to heterochromatin, an early finding that forced even Heitz to consider the possibility of heterochromatic genes (referenced in Yasuhara and Wakimoto, 2006). Drosophila heterochromatin contains "islands" of highly conserved genes embedded in these "oceans" of complex repeats, which may require special expression and splicing mechanisms (Smith et al., 2007).The existence of diverse heterochromatic genes in Drosophila did not draw sufficient interest from investigators who focused on other organisms. At first attempts, transcriptionally active genes were not found within centromeres of human and Arabidopsis sp (Hosouchi, Kumekawa, Tsuruoka, & Kotani, 2002; Schueler, Higgins, Rudd, Gustashaw, & Willard, 2001) until the end of 1990s. Copenhaver et al. (1999) reported 160 genes within the 4.3 and 2.7 Mb regions that include the centromeres of chromosomes 2 and 4 of *Arabidopsis thaliana*. At least 25 of the annotated genes are transcribed, based on cDNA evidence. A rice centromere, Cen8, which has a low content of highly repetitive CentO satellite sequences but is highly enriched for other repeated sequences contains 16 active genes (Nagaki et al., 2004). Cen8 flanking regions contain an estimated 136 active genes, with an abundance of TEs in intergenic regions (Yan et al., 2005). Recent data suggest that tomato and sorghum could have a greater fraction of genes within or near heterochromatin compared with Arabidopsis or rice (reviewed in Yasuhara & Wakimoto, 2006).

In mammals, only several transcribed genes were mapped to pericentromeric regions to date. Two genes were found in the juxtacentromeric region of chromosome 1, *BAGE2* and *TPTE*, which are transcribed only in testis and/or tumors (Brun et al., 2003; Ruault et al., 2002). *TPTE* escapes silencing in some patients with ICF syndrome (Brun et al., 2011). In chromosome 21, the genes are adjacent to the alphoid DNA array (Brun et al., 2003). Additional copy of *TPTE* is located in chromosomes 13. Two more human heterochromatic genes are transcribed in tumor or testis only: an *NF1*-related gene, which maps close to the centromere of chromosome 15, is only expressed in neuroblastoma (Legius et al., 1992) and *SLC6A10*, which maps to the juxtacentromeric regions of chromosome 16, is exclusively expressed in testis (Iyer et al., 1996). *POTE* genes are expressed in a few normal tissues (namely, testis, prostate, ovary, and placenta), in embryonic stem cell lines, and in various cancers as well as in patients with ICF syndrome (Bera et al., 2006, 2002; Brun et al., 2011). Twenty-eight genes/messenger RNAs for which there was evidence of transcription (as determined by best EST placement) were identified in human pericentromeres (5 Mb both ways from centromere) by She et al. (2004). The authors selected 16 distinct pericentromeric genes, mRNA, and/or ESTs where there was evidence of either exon fusion or exaptation for further expression analysis. Almost all assays (15 out of 16) amplified complementary DNA from the testis and more than half showed evidence of transcription (9 out of 16) from the ovary. Interestingly, seven of the assays were exclusively expressed from the testis. These data together with those cited above suggest that germline tissues are much more likely to express pericentromeric transcripts than any other human tissue.

Three genes—*ZNF11/33B, KIAA0187,* and *RET*—were mapped to the boundary between satellite-rich interchromosomally duplicated DNA and chromosome-specific DNA on chromosome 10 (Guy et al., 2000). *RET* proto-oncogene is expressed during the early development of human embryos between 23 and 42 days. The gene is expressed in the developing kidney (nephric duct, mesonephric tubules, and ureteric bud), the presumptive enteric neuroblasts of the developing enteric nervous system, cranial ganglia (VII + VIII, IX, and X), and in the presumptive motor neurons of the spinal cord (Attié-Bitach et al., 1998).

The existence of active genes that normally reside within heterochromatin has important implications for understanding how chromosomal context and chromatin structure can regulate gene expression.

4.2.2 Satellite DNA transcripts

Beside heterochromatic genes and genes fragments, satellite DNA of centromeric and pericentromeric regions are also transcribed. Sequencing of such transcripts confirmed that they consist of satellites only. They are polyadenylated (Enukashvily et al., 2007; Enukashvily, Malashicheva, & Waisertreiger, 2009; Kuznetzova et al., 2012; Rizzi et al., 2004) and have mostly intranuclear spot-like localization (Enukashvily et al., 2009; Valgardsdottir et al., 2005). The reported length of transcripts in mammals varied from 20 bp to 5 kb (Carone et al., 2009; Enukashvily et al., 2007; Lu and Gilbert, 2007; ; Ting et al., 2011; Valgardsdottir et al., 2005; Wong et al., 2007) probably because the transcripts were revealed at different stages of their processing. The detailed analysis of the pericentromeric satellite 3 transcripts from heat-shocked cells was performed by Valgardsdottir et al. (2005). In their study, a library of satellite 3 transcripts was created consisting of ~250,000 independent clones. The library was enriched for satellite 3 cDNA, but not free from other cDNAs. Twenty clones were randomly selected by colony hybridization, then sequenced and analyzed. The size of the inserts ranged from 19 to >1400 nt (not counting the polyA tail). All the cDNAs had a polyA tail, which ranged in size between 23 and 87 bp. In only few cases, this was preceded by a canonical polyA signal site. The cDNAs had the typical sequence of satellite 3 made of two types of repeats, a pentanucleotide (GGATT) and a decanucleotide (CAACCCGAGT). The ~1500 copies of the pentanucleotide occurred in tandem arrays composed of a variable number of repeats (from 1 to 39). Of these, 60% were identical to the consensus and most of the remaining 40% are one-nucleotide variations of this sequence. The most frequent variations were GCAAT (60 times), GGAAA (54 times), and GGATT, GTAAT, GGAAC, and AGAAT (43–45 times each). The pentanucleotide arrays are terminated by a decanucleotide monomer. The consensus decanucleotide was found a total of 29 times and accounted for ~25% of all decanucleotides in the clones. Several variations in the termination signal were found, usually differing from the consensus sequence for one to two nucleotides. The most frequent ones (CAACCAGAGT, CAACCCGAAT, and CGTTCCGAGT) were found four times each. The authors demonstrated after thorough analysis that "because of the variations in the penta- and decanucleotides sequences and in the number of repeats in each array, each satellite 3 transcript has a unique structure."

Taking into account the extreme sequence diversity of satellite DNAs and their transcripts, several sequence-specific regulatory signals might reside within them (reviewed by Ugarkovic, 2005). The characteristic sequence structure of some satellite DNAs is based on simple repeats, which led to the proposal that they are transcribed by read-through from upstream genes or TE promoters (Diaz et al., 1981). However, the sequence of satellite 2 found in the newts *Notophthalmus viridescens* and *Triturus vulgaris meridionalis* contains a functional analogue of the vertebrate snRNA promoter that is responsible for RNA pol II transcription (Coats et al., 1994). Human satellite III which is specifically expressed under stress has a binding motif for the heat-shock transcription factor 1 that drives RNA pol II transcription (Metz et al., 2004).

Transcription of satellites is asymmetrical (Enukashvily et al., 2007, 2009; Eymery, Horard, et al., 2009; Kuznetzova et al., 2012; Rizzi et al., 2004; Rudert et al., 1995; Valgardsdottir et al., 2005 and reviewed in Eymery, Callanan and Vourch, 2009; Vourc'h and Biamonti, 2011). In different cells, at different stages of development or cell life, either sense (T-rich in mouse, GGAAT) or antisense (A-rich in mouse, ATTCC rich) is transcribed. In cells under stress conditions, transcripts in sense orientation are usually observed (Eymery, Horard, et al., 2009). However, in heat-shocked cells, twofold decrease in the amount of antisense satellite 3 transcripts and a ninefold increase in the amount of sense satellite 3 transcripts were demonstrated by Jolly et al. (2004). In embryogenesis, the orientation of transcripts is regulated in more subtle way. At the earliest stages of mouse embryogenesis, major satellite repeats are highly expressed and the expression is strand specific. A switch between chains was observed between 1- and 2-cell stages. Later, the transcription was downregulated (Probst et al., 2010). However, it is not the only burst of strand-specific pericentromeric DNA transcription in mouse embryo development. Enukashvily et al. (2009) demonstrated strand-specific transcription of pericentromeric satellites in embryonic stem cells culture induced by retinoic acid. Rudert et al. (1995) demonstrated that in mouse embryo sense transcripts of major satellite appeared in many tissues especially in central nervous system (CNS) at 11.5–15.5 day postcoitum (dpc). Transcripts in opposite direction were observed only in a small subset of CNS cells. In adult mice testis, PCT transcripts in an antisense orientation are observed in immature germ cells, whereas in a sense orientation they are observed in mature germ cells (Rudert et al., 1995). In human, the spatial and temporal separation of pericentromeric satellite 3 transcription was also demonstrated recently (Kuznetzova et al., 2012). The transcripts were observed in embryos up to 9-weeks old. The simultaneous transcription from both chains was never observed in any of the samples analyzed. The transcripts from sense chain were found in medulla oblongata, lung, intestine, chorion of 7-week-old fetus. In heart and kidney, the transcription was going on in opposite direction.

The chain selective transcription gives a hint that it is thoroughly regulated probably by transcription factors and appears not randomly upon requirement.

4.2.3 Transcription in stress condition

One of the most well-studied examples of transcriptional activation of centromeric specific sequences is that which occurs in response to cell stress. Transcription of satellites can be induced by different factors. Oxidative stress and genotoxic agents (such as etoposide, aphidicolin) induce low-level transcription of the pericentromeric HC, whereas hyperosmotic stress and UVC induce middle-level activation of the transcription. Heat shock and cadmium are shown to be the strongest activators of satellite DNA transcription among other stress stimuli (Valgardsdottir et al., 2008 and reviewed by Eymery, Callanan, et al., 2009). In 1993, Sarge et al. demonstrated the preferential binding of HSF-1 with heterochromatic regions of 9q12 (Sarge et al. 1993). Satellite 3 of 9q12 region is a target for this interaction (Jolly et al., 2002). Two years later, the activation of 9q12 satellite 3 DNA transcription was demonstrated in heat-shock-treated cells by two research groups almost simultaneously (Jolly et al., 2004; Rizzi et al., 2004). After heat shock, the 9q12 region acquires some euchromatic features—HP1 and histone H3 methylated on lysine 9, and are excluded from heterochromatic nuclear districts (Rizzi et al., 2004). The transcripts of satellite 3 of different lengths are accumulated in this region and are necessary for nuclear stress body formation (Valgardsdottir et al., 2005). Now it is proven that cell stress induced by different stimuli is accompanied with nuclear stress body formation and pericentromeric DNA transcription (Sengupta, Parihar, & Ganesh, 2009). Within nuclear stress bodies, ncRNAs recruit a number of RNA-binding proteins involved in pre-mRNA processing, including splicing regulators (Metz et al., 2004). Originally, chromosome specificity of heat-shocked-induced transcription was assumed (Jolly et al., 2004). Later it was shown that heat shock triggers pericentromeric satellites transcription on several chromosomes in human (Eymery, Souchier, Vourc'h, & Jolly, 2010). The mechanism underlying the specificity of HSF1 binding to chromosome 9 is still unclear. Sandqvist et al. (2009) demonstrated that heterotrimerization of HSF1 and HSF2 is required for satellite 3 binding. Heterotrimerization provides a switch that integrates transcriptional activation in response to stress and developmental stimuli.

Similar activation of satellites transcription is observed in other species. In mice, heat shock (but not other types of stresses) induces transcription of *cassini* locus (a murine gamma/major satellite sequence; Arutyunyan et al., 2012). The authors found that cassini was highly upregulated in drug-treated Acute lymphoblastic leukemia (ALL) cells. Analysis of RNAs from different normal mouse tissues showed that *cassini* expression is highest in spleen and thymus, and can be further enhanced in these organs by exposure of mice to bacterial endotoxin. These data suggest that the transcripts might have protective effects in stressed cells. Pezer and Ugarkovic (2012) studied the transcription of abundant satellite DNA TCAST that makes up 35% of genome of the red flour beetle *Tribolium castaneum* and is located within the constitutive pericentromeric heterochromatin. They found that expression is strongly induced following heat shock and is accompanied by increase in repressive epigenetic modifications of histones at TCAST regions providing a toll for protection cells from heat shock.

Recently, new data were published about the involvement of Daxx in regulation and maintenance of heat-shock–induced satellite 3 transcription in human. Daxx, a Death domain-associated protein, functions as a potent transcription repressor that binds to sumoylated transcription factors. Its role in apoptosis was intensively studied, but recently a new Daxx role was unveiled as a novel chaperone for histone H3.3. In heat-shock–treated cells, Daxx leaves PML bodies (where it localizes in untreated cells) and associates with centromeric/pericentromeric domain. The burst of satellite 3 of chromosome 9 but not alphoid DNA transcription occurs in the first 2 h after heat shock. In Daxx-depleted cells, this burst is much less prominent— the number of copies is twice decreased as compare to control cells with intact Daxx (Morozov, Gavrilova, Ogryzko, & Ishov, 2012). The authors found that Daxx repression or mutations reduce incorporation of H3.3 on tandem repeats, reduce pericentromeric transcription, and change epigenetic marks in centromere/pericentromere regions.

4.2.4 Transcription in development

The accumulation of centromeric and PCT transcripts takes place in the course of embryogenesis. The periods of transcription bursts are intermingled with periods of relative transcriptional silence.

A detailed study of satellite DNA transcription in mouse preimplantation embryo was performed by Probst et al. (2010). In their study, the evidence that major satellite RNA is synthesized *de novo* from zygote genome at early preimplation studies was obtained. The transcriptional upregulation followed by a rapid downregulation coincides with the reorganization of pericentric satellites into chromocenters during a discrete time window at the 2-cell stage. The transcription is strand specific—the forward chain transcripts accumulation at 2-cell stage (late S/early G2) is preceded by the reverse chain transcripts accumulation in paternal pronucleus in zygote. This transcription is shown to be necessary for heterochromatic chromocenters formation. Santenard et al. (2010) reported the accumulation of histone 3.3 in paternal genome pericentromeric heterochromatin. Mutation of H3.3 K27 (monomethylation on lysine K27), but not of H3.1 K27, in early stages (2 cells) of mouse embryo results in aberrant accumulation of pericentromeric transcripts, HP1 mislocalization, dysfunctional chromosome segregation, and developmental arrest. This phenotype is rescued by injection of double-stranded RNA (dsRNA) derived from pericentromeric transcripts, indicating a functional link between H3.3K27 and the silencing of such regions by means of an RNA-interference (RNAi) pathway.

Burst of constitutive HC DNA transcription takes place also later in embryogenesis. In E-14 cells, a cell line originated from undifferentiated mouse embryonic stem cells, the transcription of major but not minor satellite begins next day after differentiation induction by retinoic acid. On the first day after stimulation, transcripts are localized outside chromocenters, but later from the day 2 onwards, most of the hybridization signals are adjacent to chromocenters (Enukashvily et al., 2009).

Pericentromeric HC satellite DNA transcripts are found also at the later stages of mouse embryogenesis. Rudert et al. (1995) detected the transcripts of pericentromeric satellite DNA in the whole embryo 11.5–15.5 dpc, in the CNS at 12.5 dpc and in scattered cells from the CNS at 15.5 dpc, whereas in adult tissues, they have been detected in liver and testis only.

In human, Eymery, Horard, et al. (2009) found only centromeric transcripts during screening embryo liver and placenta RNA pools purchased from Clontech. Kuznetzova et al. (2012) revealed tissue-specific and temporally specific pericentromeric transcription at 7–11 weeks of gestation at least in tested samples (Table 2.1). In this work, the pericentromeric transcripts were not observed between 11 and 13 weeks of gestation (samples from older fetuses were not available in these experiments).

Transcripts of pericentromeric satellites were also found in some adult tissues. In mouse, the transcripts were found only in liver and testis, that is, in organ and cells capable of high proliferation (Rudert et al., 1995). In human, transcripts of centromeric DNA were found in ovary, whereas pericentromeres were transcribed in testis (Eymery, Horard, et al., 2009).

Table 2.1 Transcription of pericentromeric satellite 3 in human fetuses at the age of gestation from 7 to 13 weeks (according to Kuznetzova et al., 2012)

Age of gestation (in weeks)	Organ	Pericentromeric satellite 3 transcription
7	Chorion	+
	Medulla oblongata	+
	Kidney	−
		−
9	Brain	−
	Medulla oblongata	+
	Adrenal gland	+
	Kidney	±
	Intestine	+
	Liver	+
	Heart	+
	Chorion	−
11–12	Chorion	−
	Medulla oblongata	−
	Adrenal gland	−
	Liver	−
	−	−
12–13	Chrorion	−

+, transcribed in all patients screened; −, no transcription detected in any of the patients; ±, transcribed in some patients.

Recently, Ting et al. (2011) reported alphoid DNA to be the most abundant class of normally expressed human satellites in human pancreas.

4.2.5 Cell-cycle-regulated transcription of pericentromeric satellites

Under physiological conditions, transcripts of pericentromeric satellites in somatic cells were observed at the two stages of life cycle—molecules smaller than 200 bp were detected specifically in mitotic cells and were undetectable by 1 h after mitosis, whereas molecules of larger size (>1 kb) during the course of G1 reaching a peak in late G1-/early S-phase (Lu and Gilbert, 2007). Cell cycle regulation of γ satellite transcription is independent of the Suv39h1,2-related modifications of heterochromatin. It was demonstrated that pericentromeric transcription requires activation of Cdk and passage through the checkpoints. Transcription of small centromeric (120 bp) sequences was also demonstrated (Bouzinba-Segard, Guais, & Francastel, 2006). The overaccumulation of these transcripts leads to chromosome missegregation, loss of sister-chromatid cohesion, and aneuploidy.

In human, the importance of alphoid RNA in nucleolus and centromere formation was demonstrated by Wong et al. (2007). The authors showed that CENP-C protein, one of the constitutive centromere proteins crucial for centromere assembly, is not only a DNA-binding protein (as shown earlier by (Sugimoto, Kuriyama, Shibata, & Himeno, 1997) but also capable of single-strand-RNA binding. Moreover, the binding is alphoid RNA specific and occurred *in vivo*. Single-stranded alphoid DNA was necessary for targeting CENP-C and INCENP to centromere. It is required for nucleolus localization of CENP-C and INCENP and is also an integral part of kinetochore. Inactivation of CENP-C leads to decrease in cell population growth, the mutated CENP-C with mutations in RNA-binding domain cannot restore the cell population growth. *In situ* replenishment of single-stranded α-satellite RNA restores the relocalization of CENP-C and INCENP to the centromeres of some of the chromosomes following RNase treatment. Thus, interaction between single-stranded RNA and CENP-C is necessary for kinetochore formation and cell population growth, that is, for cell proliferation. Du, Topp, and Dawe (2010) obtained similar data on maize. Single-stranded RNA transcribed from centromeres remains bound within the kinetochore region and is necessary for binding CENP-C to DNA. However, in the *in vitro* part of their experiments, the authors demonstrated that both centromeric and noncentromeric RNAs can promote binding CENP-C to maize centromere DNA.

Taken together these experiments on human, mouse, and maize cells confirm the importance of single-stranded centromeric RNA for kinetochore assembly and probably explain satellite DNA transcription during the cell cycle.

4.2.6 Transcription in senescent cells

Transcription of constitutive HC DNA occurs not only in embryogenesis, cycling cells, or some adult tissues with proliferating cells. Expression of pericentromeric polyadenylated transcripts was also observed at late passages in replicatively senescent primary fibroblasts (Enukashvily et al., 2007). Gaubatz and Cutler (1990) analyzed cells from different organs of adult and aged mice. They found pericentromeric satellite DNA transcripts only in cardiomyocytes of aged mice. The transcripts were not detected in the heart muscle of young adult animals (between 2 and 6 months), but then appeared at the age of 12 months and continued to increase over twofold up to the age of 32 months. The upregulation of pericentromeric satellite 3 transcription and loss of heterochromatic epigenetic marks (such as histone H3 trimethylated on lysine 9) were also observed in cells of patients with premature aging syndromes (Enukashvily et al., 2007; Shumaker et al., 2006).

4.2.7 Satellite DNA transcription during carcinogenesis
Under normal conditions, transcripts of satellite DNA are found in developing embryo and cycling cells of adults. The last burst of transcription occurs in some tissues. However, some pathological processes such as malignization are also accompanied with elevated satellite DNA transcription, decondensation, and demethylation (Alexiadis et al., 2007; Enukashvily et al., 2007; Ting et al., 2011; Zhu et al., 2011). Originally found in Wilms neuroblastoma tumors (Alexiadis et al., 2007) the presence of transcripts was confirmed in epithelial carcinomas cells A431 but not in HeLa (Enukashvily et al., 2007). Mutations in tumor suppressor BRCA1 that lead to breast and/ or ovarian cancer in mice also result in transcriptional derepression of the tandemly repeated satellite DNA. The elevation of major satellite transcription is greater as compare to minor satellite transcription (27-fold vs. 10-fold) Ectopic expression of satellite DNA can phenocopy BRCA1 loss in centrosome amplification, cell cycle checkpoint defects, DNA damage, and genomic instability. The global expression of repetitive ncRNAs in primary tumors was analyzed by Ting et al. (2011). The centromeric and pericentromeric DNAs were found to be greatly overexpressed in mouse and human epithelial cancers. In 8 of 10 mouse pancreatic ductal adenocarcinomas, pericentromeric satellites accounted for a mean 12% (range of 1–50%) of all cellular transcripts, a mean 40-fold increase over that in normal tissue. The authors assumed according to their data on general transcription profiles that the whole cellular transcriptional machinery is affected by the massive expression of satellites. The transcription was intensified during malignization progress in pancreatic tumors. In human, the satellite transcription was also increased in primary tumors. But some of the satellites were transcribed in normal tissues as well. The alphoid DNA was shown to be the most abundant class of normally expressed human satellites. The

most drastic changes (131-fold differential expression) were observed for satellite 2 (very similar to satellites 1 and 3) transcripts that are practically absent in normal tissues. The tumor satellite RNA revealed by Ting et al. was found in both sense and antisense directions and were absent from purified polyadenylated RNA. This finding probably suggests that the transcription of pericentromeric DNA is different in tumors from similar processes in development and cell stress.

4.2.8 Regulatory mechanisms of satellite DNA transcription

Transcription of satellite DNA is regulated by different pathways in different cells and conditions. The very first know satellite transcription regulator to control the expression of PCT sequences was the retinoic acid receptor (RAR) (Rudert et al., 1995). Its effect on pericentromeric DNA transcription was different in different cell models—it can act both ways (depress or rerepress) depending on cell types (Martens et al., 2005; Rudert et al., 1995) and probably on the presence or absence of binding sites of RAR in mouse pericentromeric satellites (Martens et al., 2005). Some of the sequences capable of binding to RAR were identified in centromeric heterochromatin—for example, some of Alu-repeats and some sites within major satellites itself (Rudert et al., 1995; Rudert & Gronemeyer, 1993; Vansant & Reynolds, 1995). Pericentromeric but not centromeric satellites are known to be associated with Ddx 5 protein (RNA-helicase p68) in differentiating mouse E-14 embryonic stem cells as well as in human liver of third trimester fetuses and mouse adult liver (Enukashvily, Donev, Sheer, & Podgornaya, 2005; Enukashvily et al., 2009). In differentiating E-14 cells, the interaction occurs only after stimulation with retinoic acid. One of the Ddx5 functions is regulation of transcription (reviewed by Fuller-Pace and Ali, 2008). It suggests that the satellite DNA-Ddx5 interaction might play a role in transcription regulation during embryogenesis and differentiation.

The transcripts revealed in primary tumors are different from those revealed in normal tissues—they are of larger size and not polyadenylated (Ting et al., 2011). This suggests that their transcription is regulated in different ways and probably occurs at different sites of satellite DNA. In cancer, satellites are shown to be decondensed and co-transcribed with autonomous retrotrasposones such as LINE-1 (Ting et al., 2011). LINE-1 insertion upstream of transcriptional start sites of cellular transcripts has recently been implicated in gene regulation; therefore, LINE insertions that are present in perocentromeric arrays might be involved satellite DNA transcription regulation during carcinogenesis.

It has been proposed that retroelements may facilitate or initiate the satellite sequence transcription observed in normal tissues of a broad range of eukaryotic species (Ugarkovic, 2005). Carone et al. (2009) uncovered the strong bidirectional promoter capability of the kangaroo endogenous retrovirus (KERV-1) LTR to produce long double-stranded RNAs for both KERV-1 and surrounding sequences, including sat23. These long dsRNAs are then processed by an unknown mechanism into centromere repeat associated small interacting RNAs (crasiRNAs), 34–42 nt in length. The crasiRNAs are involved in the recruitment of heterochromatin and/or centromeric proteins. These small centromere-associated ncRNAs occur conserved among eukaryotes suggesting their impact also in human. The authors propose that production of small RNAs from transcription of centromeric repeats driven by retroviral promoters is a conserved feature of eukaryotic centromeres. Komissarov et al. (2011) described the presence of MTA transposons similar to ERV-3 retrovirus in pericentromeric regions of mouse chromosomes. This finding supports the hypothesis made by Carone et al. (2009).

Unlike the transcription in cancer, development and proliferation, the transcription of satellites activated by heat shock is chromosome specific. Three transcription factors have been identified so far that control the expression of pericentromeric satellite 3 sequences in human cells in response to stress—HSF1, HSF2, and TonEBP (reviewed by Vourc'h and Biamonti, 2011), another transcription factor YY1 associates with pericentromeric murine DNA in proliferating but not quiescent cells (Shestakova, Mansuroglu, Mokrani, Ghinea, & Bonnefoy, 2004). The first two factors are involved in cellular response to heat shock, TonEBP is activated in response to high concentrations of osmolytes. All of these factors relocate to the nuclear stress bodies upon stress activation (Jolly et al., 2004; Metz et al., 2004; Vourc'h and Biamonti, 2011), HSF proteins are the main regulators of HSP synthesis (reviewed by Sreedhar and Csermely, 2004) and TonEBP rather than HSF stimulates transcription of the HSP70-2 gene in response to hypertonicity (Woo, Lee, Na, Park, & Kwon, 2002). The involvement of YY1 in satellite DNA transcription regulation is not shown yet. Canonical HSF1 binding sites are not present in the prototypical Satellite 3 element in pHuR98 plasmid (accession number GenBank: X06137.1). However, in an *in vitro* EMSA assay, this sequence is specifically recognized and bound by HSF1. HSF2 factor is required for HSF1–HSF2 heterotrymerization, that in turn is a necessary stage in binding HSF1-HSF2 to pericentromeric satellites (Sandqvist et al., 2009; Vourc'h and Biamonti, 2011). Even less is known in the case of TonEBP although *in silico* analysis of the satellite 3 sequence picked up motifs matching the consensus binding site (TGGAAANN(C/T)N(C/T)) of this factor (Valgardsdottir et al., 2008).

Two transcription factors, Pax3 and Pax9, were shown recently to be involved in major satellites transcription repression and heterochromatin formation (Bulut-Karslioglu et al., 2012). Single knockdown of Pax3 or Pax9 in wild-type mouse embryonic fibroblasts resulted in an increase in major satellite transcription—of around 10-fold for shPax9 and 18-fold for shPax3. Thus, Pax3 and Pax9 are not only redundant but also synergistic in safeguarding heterochromatin silencing. The intact sites within major satellites for binding Pax3 and Pax9 are necessary for histone H3 methylation at lysine 9. Additionally, bioinformatic screening of all histone methyl-transferase Suv39h-dependent heterochromatic repeat regions in the mouse genome revealed a high concordance with the presence of transcription factor binding sites. The authors suggest a pathway in which transcriptional repression of pericentric repeats by sequence-specific transcription factors is essential for the integrity of heterochromatin, considerably expanding the role of transcription factors beyond euchromatic gene regulation. This function is not restricted to Pax3 and Pax9, as mouse major satellite repeats contain multiple binding sites for many other transcription factors (Bulut-Karslioglu et al., 2012).

The most well-described factors that influence heterochromatic genes and satellites transcription are changes in epigenetic marks such as DNA and histone methylation, histone acetilation, and histone variants (Almouzni and Probst, 2011; Eymery, Horard, et al., 2009; Vourc'h and Biamonti, 2011). The connection between satellites transcription, degree of condensation, and epigenetic modifications was reported by many authors. Current evidence indicates that both DNA methylation and histone H3 lysine 9 (H3K9) trimethylation are critical for the maintenance of satellite repression. Targeting of DNMT3B methyltransferase (one of the three methyl transferases responsible for genomic DNA methylation) is probably mediated by CENP-C protein (Gopalakrishnan, Sullivan, Trazzi, Della Valle, & Robertson, 2009). The influence of demethylating agents (Haaf and Schmid, 2000) and transcription inhibitors (Haaf and Ward, 1996) on constitutive HC condensation in pericentromeric regions was demonstrated in human lymphocytes. In senescent cultures of fibroblasts, the methylation level of satellite 3 is decreased twice faster than in the rest of genome (Suzuki et al., 2002) and this rapid demethylation is accompanied by 1q12 decondensation and its satellite 3 transcription (Enukashvily et al., 2007). The selective demethylation of satellites 1 and 2 in cancer cells was demonstrated in several studies (Ehrlich et al., 2003; Jackson et al., 2003; Narayan et al., 1998; Qu, Grundy, Narayan, & Ehrlich, 1999) and hypomethylation of pericentromeric satellite is one of the initial stages of cancerogenesis in breast cancers (Jackson et al., 2003). The undermethylation of pericentromeric regions is accompanied with the transcription of heterochromatic genes in patients with ICF syndrome (Brun et al., 2011). Sugimura et al. (2010) demonstrated that reduced level of DNA methylation was associated with the pericentromeric HC enrichment in euchromatic histone modifications (acetylation of histone H4, and di- and trimethylation of lysine 4 on histone H3) and with the activation of transcription from centromeric and pericentromeric satellite repeats. However, some authors argue against the crucial role of DNA methylation in regulation of satellite DNA transcription (Eymery, Callanan, et al., 2009, Eymery, Horard, et al., 2009). The undermethylation of tandemly repeated DNA might influence indirectly the transcription profile of the whole genome: given the highly reiterated nature of satellite DNA, these sequences have high potential to act as holding or delivery docks for transcription control proteins (Ehrlich, 2009).

The histone methylation (H3-K9me3, H4K20me3, and H3K27me3), histone deacetilation, and histone vartiants are also involved in regulation of heterochromatin DNA transcription (Bulut-Karslioglu et al., 2012; Lu & Gilbert, 2007; and reviewed in Almouzni and Probst, 2011; Vourc'h and Biamonti, 2011). Centromeric chromatin in humans contains interspersed blocks of CENP-A (a histone H3 variant) and H3-containing nucleosomes (reviewed Choo, 1997; Schueler and Sullivan, 2006). Human centromeric chromatin is flanked by pericentromeric heterochromatin enriched in repressive H3K9 di- and trimethylation. Centromeric H3 nucleosomes are marked by H3K4 dimethylation (transcriptional permissive mark) and by a lack of H3 and H4 acetylation (hypoacetylation is a hallmark of heterochromatin). Thus, centromeres have their own epigenetic marks pattern (Sullivan and Karpen, 2004). Other studies have revealed enrichment of H3K9 dimethylation at α-satellite DNA and to a lesser extent H3K9 trimethylation and H3K27 mono- and trimethylation. H3K9 dimethylation may act as a boundary between centromeric chromatin containing CENP-A and pericentromeric heterochromatin containing H3K9 trimethylation. These findings suggest that a balance between repressive heterochromatin and CENP-A-containing heterochromatin is essential for proper centromere function (referenced in Gopalakrishnan et al., 2009). Hystone H3 methylation on lysine 9 is necessary for HP1 binding, heterochromatin silencing and depends on DNA methylation (referenced in Almouzni and Probst, 2011). In mouse embryos, the low levels of K27me and the presence of H3.3 provide an opportunity for heterochromatic repeats transcription in the male pronucleus (Santenard et al., 2010).

The epigenetic marks are interconnected. Histone modifications depend on DNA methylation and vice versa. The inhibition of DNA methylation leads to heterochromatin DNA transcription derepression and pericentromeric heterochromatin enrichment in euchromatic histone modifications such as acetylation of histone H4, and di- and trimethylation of lysine 4 on histone H3. Histone H3.3 (a mark of active genes) moved to the pericentromeric heterochromatin prior to the accumulation of the other euchromatic histone modifications (Sugimura et al., 2010).

5. FUNCTIONS OF CENTROMERIC AND PERICENTROMERIC HC DNA TRANSCRIPTION

Two types of sequences are transcribed from centromeric and pericentromeric regions in different species—heterochromatic genes and non-coding DNA of different nature (retrotransposones, genes fragments, and tandemly repeated satellite DNA). The issue that whether this transcription is functional or is just a consequence of genome destabilization and disruption of heterochromatin is still discussed. However, now it becomes more and more obvious that the transcription of heterochromatic DNA is an important factor in genome functioning providing a tool for genome structural organization.

In human, the transcription of heterochromatic genes *TPTE, TPTE2, POTE, BAGE* is normally observed in testis and ovaries, but it is reactivated during cancerogenesis and in patients with ICF syndrome (for references see Brun et al., 2011). It seems logical to suggest that the transcription of heterochromatic genes is involved in developing the ICF syndrome phenotype, but there is no direct proof of the hypothesis yet. Another gene juxtaposing the satellite-rich heterochromatin on chromosome 10—*RET* plays a major role in development of mouse enteric nervous system and in kidney organogenesis.

Pericentromeric satellite DNA transcription is probably multifunctional. One of the most well-documented functions is involvement in heterochromatin formation and maintenance. The inactivation of pericentromeric transcripts during the 2-cell stage results in embryos that fail to form chromocenters and arrest their development in the G2-phase of the second cell cycle (Probst et al., 2010). The role of mammalian pericentromeric satellites in chromatin formation is probably similar to S. pombe. In S. pombe, 20–30 bp transcripts of PCT origin are generated by the RNAi machinery and targeted to pericentromeric regions. Two RNAi complexes, the RNA-Induced Transcriptional Silencing complex (RITS), which contains an siRNA bound to an Argonaute protein, and the RNA-Directed RNA polymerase Complex (RDRC), are critical components to the deposition of H3K9me and heterochromatin marks (Grewal and Elgin, 2002; Vourc'h and Biamonti, 2011). In chicken, downregulation of Dicer leads to accumulation of short (20–30 bp) pericentromeric transcripts and mitosis abnormalities occur due to premature sister-chromatid separation rather than centromere impair (Fukagawa et al., 2004). Though some data argued against the importance of RNAi processing of pericentromeric transcripts role in mammalian heterochromatin maintenance (Cobb et al., 2005; Korhonen et al., 2011), recent publications give a hint that satellite DNA transcripts are involved in similar processes in higher eukaryotes (Bouzinba-Segard et al., 2006; Lu and Gilbert, 2007; Valgardsdottir et al., 2005). In mammals, the processing of pericentromeric transcripts by RNAi machinery is probably restricted to undifferentiated embryonic stem cells and germ cells (Kanellopoulou et al., 2005; Korhonen et al., 2011) when genome heterochromatin is formed *de novo*. In these cells, downregulation of Dicer leads to the accumulation of long satellite transcripts (Kanellopoulou et al., 2005; Korhonen et al., 2011). However, direct involvement of RNAi processing in centromeric transcripts processing and silencing is yet to be proven.

Satellite transcripts are involved in kinetochore formation and centromere assembly where they bind CENP-C centromeric protein and are probably responsible for targeting the protein to kinetochore region (Wong et al., 2007). The association of the accumulation of full-length centromere satellite transcripts with centromere failure in stressed cells (Bouzinba-Segard et al., 2006; Valgardsdottir et al., 2005) argues that these transcripts must be processed in some way to function properly. Carone et al. (2009) demonstrated the association of transcripts, consisting of satellites and a retrovirus, with centromere in tammar wallaby. These RNAs are a product of processing of full-length retroposon-satellite transcripts, but the mechanism of processing is unknown yet. The authors suggest the existence of yet unknown mechanism for processing long satellite RNAs into short fragments that bind centromeres.

Heterochromatin transcription can also be utilized by a cell as a tool for sequestrating transcription factors due to highly repetitive character of satellite arrays. For example, massive recruitment of transcription and splicing factors into nuclear stress bodies after heat-shock might be a tool for sequestration of pre-mRNA processing factors (reviewed in Jolly & Lakhotia, 2006).

An interesting hypothesis was published by Parris (2010). It is suggested that "development of multi-cellular organisms is guided by a Master Development Program (MDP) located primarily in the pericentromeric heterochromatin. The MDP is believed to consist of a series of Generation-Specific Control Keys (GSCK) transcribed in sequence by Ikaros family transcription factors unless the GSCKs are suppressed by Sall1-family or Dnmt3b-family proteins. The MDP is proposed to increment with each cell cycle to the next GSCK resulting in development of the clone. The transcripts of the GSCKs presumably yield noncoding nuclear messenger RNAs (nmRNAs, 8–30 nt units) that act directly (e.g., as primers for RNA polymerase II) and indirectly to regulate HOX and other high-level transcription factor and developmental genes. As envisioned, the MDP would evolve by terminal addition of new GSCKs. The new GSCKs are produced by evolutionary consolidation of retro-transcripts into pyknons that collect and evolve at the end of the pericentromeric heterochromatin and are eventually incorporated into the MDP." Whether this hypothesis is true or false remains to be further investigated.

6. CONCLUSIONS

Thus, our ideas about the role of non-coding centromeric DNA made a leap from "junk DNA" to "Generation Specific Control Keys." As usual the truth is something in between these two hypothesis that represents two poles of our knowledge about centromere/pericentromere functions in a living cell.

Many questions are to be answered. The first one is, what are the exact functions of satellite DNA transcripts in cells of different tissues in different species? Are these functions ancient and highly conserved or they vary from tissue to tissue or from species to species? How is this transcription regulated? What we know now is only a tip of an iceberg. The future investigations will give us answer about the role of noncoding DNA in a genome orchestra.

ACKNOWLEDGMENTS

The work was supported by MCB grant from Presidium of Russian Academy of Sciences (N01200955639) and RFBR (N 11-04-01676 and 11-04-01639).

REFERENCES

Alazami, A. M., Mejia, J. E., & Monaco, Z. L. (2004). Human artificial chromosomes containing chromosome 17 alphoid DNA maintain an active centromere in murine cells but are not stable. *Genomics, 83,* 844–851.

Alexiadis, V., Ballestas, M. E., Sanchez, C., Winokur, S., Vedanarayanan, V., Warren, M., et al. (2007). RNAPol-ChIP analysis of transcription from FSHD-linked tandem repeats and satellite DNA. *Biochimica et Biophysica Acta, 1769*(1), 29–40.

Almouzni, G., & Probst, A. V. (2011). Heterochromatin maintenance and establishment: Lessons from the mouse pericentromere. *The Nucleus, 2*(5), 332–338.

Amor, D. J., & Choo, K. H. (2002). Neocentromeres: Role in human disease, evolution, and centromere study. *American Journal of Human Genetics, 71*(4), 695–714.

Arutyunyan, A., Stoddart, S., Yi, S. J., Fei, F., Lim, M., Groffen, P., et al. (2012). Expression of cassini, a murine gamma-satellite sequence conserved in evolution, is regulated in normal and malignant hematopoietic cells. *BMC Genomics, 13,* 418.

Attié-Bitach, T., Abitbol, M., Gérard, M., Delezoide, A. L., Augé, J., Pelet, A., et al. (1998). Expression of the RET proto-oncogene in human embryos. *American Journal of Medical Genetics, 80*(5), 481–486.

Avivi, Y., Morad, V., Ben-Meir, H., Zhao, J., Kashkush, K., Tzfira, T., et al. (2004). Reorganization of specific chromosomal domains and activation of silent genes in plant cells acquiring pluripotentiality. *Developmental Dynamics, 230*(1), 12–22.

Baker, W. K. (1968). Position-effect variegation. *Advances in Genetics, 14,* 133–169.

Basu, J., & Willard, H. F. (2005). Artificial and engineered chromosomes: Non-integrating vectors for gene therapy. *Trends in Molecular Medicine, 11*(5), 251–258.

Bera, T. K., Saint Fleur, A., Lee, Y., Kydd, A., Hahn, Y., et al. (2006). POTE paralogs are induced and differentially expressed in many cancers. *Cancer Research, 66,* 52–56.

Bera, T. K., Zimonjic, D. B., Popescu, N. C., Sathyanarayana, B. K., Kumar, V., Lee, B., et al. (2002). POTE, a highly homologous gene family located on numerous chromosomes and expressed in prostate, ovary, testis, placenta, and prostate cancer. *Proceedings of the National Academy of Sciences of the United States of America, 99*(26), 16975–16980.

Bernard, P., & Allshire, R. (2002). Centromeres become unstuck without heterochromatin. *Trends in Cell Biology, 12*(9), 419–424.

Bonaccorsi, S., Gatti, M., Pisano, C., & Lohe, A. (1990). Transcription of a satellite DNA on two Y chromosome loops of *Drosophila melanogaster. Chromosoma, 99*(4), 260–266.

Bouzinba-Segard, H., Guais, A., & Francastel, C. (2006). Accumulation of small murine minor satellite transcripts leads to impaired centromeric architecture and function. *Proceedings of the National Academy of Sciences of the United States of America, 103,* 8709–8714.

Brun, M. E., Lana, E., Rivals, I., Lefranc, G., Sarda, P., Claustres, M., et al. (2011). Heterochromatic genes undergo epigenetic changes and escape silencing in immunodeficiency, centromeric instability, facial anomalies (ICF) syndrome. *PLoS One, 6*(4), e19464.

Brun, M. E., Ruault, M., Ventura, M., Roizès, G., & De Sario, A. (2003). Juxtacentromeric region of human chromosome 21: A boundary between centromeric heterochromatin and euchromatic chromosome arms. *Gene, 312,* 41–50.

Bulut-Karslioglu, A., Perrera, V., Scaranaro, M., de la Rosa-Velazquez, I. A., van de Nobelen, S., Shukeir, N., et al. (2012). A transcription factor-based mechanism for mouse heterochromatin formation. *Nature Structural and Molecular Biology, 19*(10), 1023–1030.

Campos, S. G. P., Rodrigues, V. L. C. C., Wada, C. Y., & Mello, M. L. S. (2002). Effect of sequential cold shocks on survival and molting rate in *Triatoma infestans* (Hemiptera, Reduviidae). *Memórias do Instituto Oswaldo Cruz, 97,* 579–582.

Carone, D. M., Longo, M. S., Ferreri, G. C., Hall, L., Harris, M., Shook, N., et al. (2009). A new class of retroviral and satellite encoded small RNAs emanates from mammalian centromeres. *Chromosoma, 118*(1), 113–125.

Choo, K. H. (1997). *The centromere*. Oxford: Oxford University Press.

Coats, S. R., Zhang, Y., & Epstein, L. M. (1994). Transcription of satellite 2 DNA from the newt is driven by a snRNA type of promoter. *Nucleic Acids Research, 22*(22), 4697–4704.

Cobb, B. S., Nesterova, T. B., Thompson, E., Hertweck, A., O'Connor, E., Godwin, J., et al. (2005). T cell lineage choice and differentiation in the absence of the RNase III enzyme Dicer. *The Journal of Experimental Medicine, 201*(9), 1367–1373.

Cohen, A. K., Huh, T. Y., & Helleiner, C. W. (1973). Transcription of satellite DNA in mouse L-cells. *Canadian Journal of Biochemistry, 51*(5), 529–532.

Cohen, A. K., Rode, H. N., & Helleiner, C. W. (1972). The time of synthesis of satellite DNA in mouse cells (L cells). *Canadian Journal of Biochemistry, 50*(2), 229–231.

Cooper, K. (1959). Cytogenetic analysis of major heterochromatic elements (especially Xh and Y) in *Drosophila melanogaster*, and the theory of "heterochromatin" *Chromosoma, 10*, 535–588.

Copenhaver, G. P., Nickel, K., Kuromori, T., Benito, M. I., Kaul, S., Lin, X., et al. (1999). Genetic definition and sequence analysis of Arabidopsis centromeres. *Science, 286*, 2468–2474.

Craig, J. M. (2005). Heterochromatin—Many flavours, common themes. *Bioessays, 27*(1), 17–28.

Dantas, M. M., & Mello, M. L. S. (1992). Changes in the nuclear phenotypes of *Triatoma infestans* Klug, induced by thermal shocks. *Revista Brasileira Genética, 15*, 509–519.

De Cecco, M., Criscione, S. W., Peckham, E. J., Hillenmeyer, S., Hamm, E. A., Manivannan, J., et al. (2013). Genomes of replicatively senescent cells undergo global epigenetic changes leading to gene silencing and activation of transposable elements. *Aging Cell.* http://dx.doi.org/10.1111/acel.12047. [Epub ahead of print].

Diaz, M. O., Barsacchi-Pilone, G., Mahon, K. A., & Gall, J. G. (1981). Transcripts from both strands of a satellite DNA occur on lampbrush chromosome loops of the newt Notophthalmus. *Cell, 24*(3), 649–659.

Dillon, N. (2004). Heterochromatin structure and function. *Biology of the Cell, 96*(8), 631–637.

Donahue, R. P., Bias, W. B., Renwick, J. H., & McKusick, V. A. (1968). Probable assignment of the Duffy blood group locus to chromosome 1 in man. *Proceedings of the National Academy of Sciences of the United States of America, 61*(3), 949–955.

Du, Y., Topp, C. N., & Dawe, R. K. (2010). DNA binding of centromere protein C (CENPC) is stabilized by single-stranded RNA. *PLoS Genetics, 6*(2), e1000835.

Dunham, I., Shimizu, N., Roe, B. A., Chissoe, S., Hunt, A. R., Collins, J. E., et al. (1999). The DNA sequence of human chromosome 22. *Nature, 402*(6761), 489–495.

Durand-Dubief, M. I., & Ekwall, K. (2008). Heterochromatin tells CENP-A where to go. *Bioessays, 30*, 526–529.

Ehrlich, M. (2009). DNA hypomethylation in cancer cells. *Epigenomics, 1*(2), 239–259.

Ehrlich, M., Hopkins, N. E., Jiang, G., Dome, J. S., Yu, M. C., Woods, C. B., et al. (2003). Satellite DNA hypomethylation in karyotyped Wilms tumors. *Cancer Genetics and Cytogenetics, 141*(2), 97–105.

Enukashvily, N. I., Donev, R. M., Sheer, D., & Podgornaya, O. I. (2005). Satellite DNA binding and cellular localization of RNA helicase P68. *Journal of Cell Science, 118*, 611–622.

Enukashvily, N. I., Donev, R., Waisertreiger, I. S.-R., & Podgornaya, O. I. (2007). Human chromosome 1 satellite 3 DNA is decondensed, demethylated and transcribed in senescent cells and in A431 epithelial carcinoma cells. *Cytogenetic and Genome Research, 118*, 42–54.

Enukashvily, N. I., Malashicheva, A. B., & Waisertreiger, I. S. (2009). Satellite DNA spatial localization and transcriptional activity in mouse embryonic E-14 and IOUD2 stem cells. *Cytogenetic and Genome Research, 124*(3–4), 277–287.

Eymery, A., Callanan, M., & Vourch, C. (2009). The secret message of heterochromatin: New insights into the mechanisms and function of centromeric and pericentric repeat sequence transcription. *International Journal of Developmental Biology, 53*, 259–268.

Eymery, A., Horard, B., El Atifi-Borel, M., Fourel, G., Berger, F., Vitte, A. L., et al. (2009). A transcriptomic analysis of human centromeric and pericentric sequences in normal and tumor cells. *Nucleic Acids Research, 37*(19), 6340–6354.

Eymery, A., Souchier, C., Vourc'h, C., & Jolly, C. (2010). Heat shock factor 1 binds to and transcribes satellite II and III sequences at several pericentromeric regions in heat-shocked cells. *Experimental Cell Research, 316*, 1845–1855.

Fazzari, M. J., & Greally, J. M. (2004). Epigenomics: Beyond CpG islands. *Nature Reviews. Genetics, 5*, 446–455.

Folco, H. D., Pidoux, A. L., Urano, T., & Allshire, R. C. (2008). Heterochromatin and RNAi are required to establish CENP-A chromatin at centromeres. *Science, 319*(5859), 94–97.

Fukagawa, T., Nogami, M., Yoshikawa, M., Ikeno, M., Okazaki, T., Takami, Y., et al. (2004). Dicer is essential for formation of the heterochromatin structure in vertebrate cells. *Nature Cell Biology, 6*, 784–791.

Fuller-Pace, F. V., & Ali, S. (2008). The DEAD box RNA helicases p68 (Ddx5) and p72 (Ddx17): Novel transcriptional co-regulators. *Biochemical Society Transactions, 36*(Pt 4), 609–612.

Gaubatz, J. W., & Cutler, R. G. (1990). Mouse satellite DNA is transcribed in senescent cardiac muscle. *Journal of Biological Chemistry, 265*(29), 17753–17758.

Gilbert, D. M. (2002). Replication timing and transcriptional control: Beyond cause and effect. *Current Opinion in Cell Biology, 14*(3), 377–383.

Gopalakrishnan, S., Sullivan, B. A., Trazzi, S., Della Valle, G., & Robertson, K. D. (2009). DNMT3B interacts with constitutive centromere protein CENP-C to modulate DNA methylation and the histone code at centromeric regions. *Human Molecular Genetics, 18*(17), 3178–3193.

Grewal, S. I., & Elgin, S. C. (2002). Heterochromatin: new possibilities for the inheritance of structure. *Current Opinion in Genetics and Development, 12*(2), 178–187.

Grewal, S. I., & Elgin, S. C. (2007). Transcription and RNA interference in the formation of heterochromatin. *Nature, 447*, 399–406.

Guenatri, M., Bailly, D., Maison, C., & Almouzni, G. (2004). Mouse centric and pericentric satellite repeats form distinct functional heterochromatin. *The Journal of Cell Biology, 166* (4), 493–505.

Guy, J., Spalluto, C., McMurray, A., Hearn, T., Crosier, M., Viggiano, L., et al. (2000). Genomic sequence and transcriptional profile of the boundary between pericentromeric satellites and genes on human chromosome arm 10q. *Human Molecular Genetics, 9*(13), 2029–2042.

Haaf, T., & Schmid, M. (2000). Experimental condensation inhibition in constitutive and facultative heterochromatin of mammalian chromosomes. *Cytogenetics and Cell Genetics, 91*(1–4), 113–123.

Haaf, T., & Ward, D. C. (1996). Inhibition of RNA polymerase II transcription causes chromatin decondensation, loss of nucleolar structure, and dispersion of chromosomal domains. *Experimental Cell Research, 224*(1), 163–173.

Harel, J., Hanania, N., Tapiero, H., & Harel, L. (1968). RNA replication by nuclear satellite DNA in different mouse cells. *Biochemical and Biophysical Research Communications, 33*(4), 696–701.

He, D., & Brinkley, B. R. (1996). Structure and dynamic organization of centromeres/prekinetochores in the nucleus of mammalian cells. *Journal of Cell Science, 109*(Pt 11), 2693–2704.

Heitz, E. (1928). Das heterochromatin der moose. *Jahrbuch der Wissenschaftlichen Botanik, 69*, 762–818.

Henning, K. A., Novotny, E. A., Compton, S. T., Guan, X. Y., Liu, P. P., & Ashlock, M. A. (1999). Human artificial chromosomes generated by modification of a yeast artificial chromosome containing both human alpha satellite and single-copy DNA sequences. *Proceedings of the National Academy of Sciences of the United States of America, 9*, 592–597.

Horvath, J. E., Viggiano, L., Loftus, B. J., Adams, M. D., Archidiacono, N., Rocchi, M., et al. (2000). Molecular structure and evolution of an alpha satellite/non-alpha satellite junction at 16p11. *Human Molecular Genetics, 9*(1), 113–123.

Hosouchi, T., Kumekawa, N., Tsuruoka, H., & Kotani, H. (2002). Physical map-based sizes of the centromeric regions of Arabidopsis thaliana chromosomes 1, 2, and 3. *DNA Research, 9*(4), 117–121.

Howe, M., Dimitri, P., Berloco, M., & Wakimoto, B. T. (1995). Cis-effects of heterochromatin on heterochromatic and euchromatic gene activity in *Drosophila melanogaster*. *Genetics, 140*(3), 1033–1045.

Ikeno, M., Grimes, B., Okazaki, T., Nakano, M., Saitoh, K., Hoshino, H., et al. (1998). Construction of YAC-based mammalian artificial chromosomes. *Nature Biotechnology, 16*, 431–439.

Iyer, G. S., Krahe, R., Goodwin, L. A., Doggett, N. A., Siciliano, M. J., Funanage, V. L., et al. (1996). Identification of a testis-expressed creatine transporter gene at 16p11.2 and confirmation of the X-linked locus to Xq28. *Genomics, 34*(1), 143–146.

Jackson, K., Yu, M. C., Arakawa, K., Fiala, E., Youn, B., Fiegl, H., et al. (2003). DNA hypomethylation is prevalent even in low-grade breast cancers. *Cancer Biology & Therapy, 3*(12), 1225–1231.

Jamrich, M., Warrior, R., Steele, R., & Gall, J. G. (1983). Transcription of repetitive sequences on Xenopus lampbrush chromosomes. *Proceedings of the National Academy of Sciences of the United States of America, 80*(11), 3364–3367.

Jolly, C., Konecny, L., Grady, D. L., Kutskova, Y. A., Cotto, J. J., Morimoto, R. I., et al. (2002). In vivo binding of active heat shock transcription factor 1 to human chromosome 9 heterochromatin during stress. *The Journal of Cell Biology, 156*(5), 775–781.

Jolly, C., & Lakhotia, S. C. (2006). Human sat III and Drosophila hsr omega transcripts: A common paradigm for regulation of nuclear RNA processing in stressed cells. *Nucleic Acids Research, 34*(19), 5508–5514.

Jolly, C., Metz, A., Govin, J., Vigneron, M., Turner, B. M., Khochbin, S., et al. (2004). Stress-induced transcription of satellite III repeats. *The Journal of Cell Biology, 164*(1), 25–33.

Kanellopoulou, C., Muljo, S. A., Kung, A. L., Ganesan, S., Drapkin, R., Jenuwein, T., et al. (2005). Dicer-deficient mouse embryonic stem cells are defective in differentiation and centromeric silencing. *Genes & Development, 19*(4), 489–501.

Kellum, R. (2003). HP1 complexes and heterochromatin assembly. *Current Topics in Microbiology and Immunology, 274*, 53–77. Review.

Komissarov, A. S., Gavrilova, E. V., Demin, S. J., Ishov, A. M., & Podgornaya, O. I. (2011). Tandemly repeated DNA families in the mouse genome. *BMC Genomics, 12*, 531.

Korhonen, H. M., Meikar, O., Yadav, R. P., Papaioannou, M. D., Romero, Y., Da Ros, M., et al. (2011). Dicer is required for haploid male germ cell differentiation in mice. *PLoS One, 6*(9), e24821.

Krasikova, A. V., Vasilevskaia, E. V., & Gaginskaia, E. R. (2010). Chicken lampbrush chromosomes: Transcription of tandemly repetitive DNA sequences. *Genetika, 46*(10), 1329–1334.

Kuznetsova, I. S., Enukashvily, N. I., Noniashvili, E. M., Shatrova, A. N., Aksenov, N. D., Zenin, V. V., et al. (2007). Evidence for the existence of satellite DNA-containing connection between metaphase chromosomes. *Journal of Cellular Biochemistry, 101*, 1046–1061.

Kuznetsova, I., Podgornaya, O., & Ferguson-Smith, M. A. (2006). High-resolution organization of mouse centromeric and pericentromeric DNA. *Cytogenetic and Genome Research*, *112*, 248–255.

Kuznetsova, I. S., Prusov, A. N., Enukashvily, N. I., & Podgornaya, O. I. (2005). New types of mouse centromeric satellite DNAs. *Chromosome Research*, *13*(1), 9–25.

Kuznetzova, T. V., Enukashvily, N. I., Trofimova, I. L., Gorbunova, A., Vashukova, E. S., & Baranov, V. S. (2012). Localisation and transcription of human chromosome 1 pericentromeric heterochromatin in embryonic and extraembryonic tissues. *Medical Genetics*, *11*(4), 19–24.

Lee, H. R., Neumann, P., Macas, J., & Jiang, J. (2006). Chromosomal localization, copy number assessment, and transcriptional status of BamHI repeat fractions in water buffalo Bubalus bubalis. *Molecular and Biological Evolution*, *23*(12), 2505–2520.

Lee, C., Wevrick, R., Fisher, R. B., Ferguson-Smith, M. A., & Lin, C. C. (1997). Human centromeric DNAs. *Human Genetics*, *100*(3–4), 291–304.

Legius, E., Marchuk, D. A., Hall, B. K., Andersen, L. B., Wallace, M. R., Collins, F. S., et al. (1992). NF1-related locus on chromosome 15. *Genomics*, *13*(4), 1316–1318.

Lehnertz, B., Ueda, Y., Derijck, A. A., Braunschweig, U., Perez-Burgos, L., Kubicek, S., et al. (2003). Suv39h-mediated histone H3 lysine 9 methylation directs DNA methylation to major satellite repeats at pericentric heterochromatin. *Current Biology*, *13*(14), 1192–1200.

Lin, C. C., Sasi, R., Lee, C., Fan, Y. S., & Court, D. (1993). Isolation and identification of a novel tandemly repeated DNA sequence in the centromeric region of human chromosome 8. *Chromosoma*, *102*(5), 333–339.

Lohe, A. R., & Hilliker, A. J. (1995). Return of the H-word (heterochromatin). *Current Opinion in Genetics and Development*, *5*(6), 746–755.

Lorite, P., Renault, S., Rouleux-Bonnin, F., Bigot, S., Periquet, G., & Palomeque, T. (2002). Genomic organization and transcription of satellite DNA in the ant *Aphaenogaster subterranea* (Hymenoptera, Formicidae). *Genome*, *45*(4), 609–616.

Lu, J., & Gilbert, D. M. (2007). Proliferation-dependent and cell cycle regulated transcription of mouse pericentric heterochromatin. *The Journal of Cell Biology*, *179*, 411–421.

Lyon, M. F. (1961). Gene action in the X-chromosome of the mouse (*Mus musculus* L.). *Nature*, *190*, 372–373.

Macgregor, H. C. (1979). In situ hybridization of highly repetitive DNA to chromosomes of *Triturus cristatus*. *Chromosoma*, *71*(1), 57–64.

Maison, C., Bailly, D., Peters, A. H., Quivy, J. P., Roche, D., Taddei, A., et al. (2002). Higher-order structure in pericentric heterochromatin involves a distinct pattern of histone modification and an RNA component. *Nature Genetics*, *30*, 329–334.

Martens, J. H., O'Sullivan, R. J., Braunschweig, U., Opravil, S., Radolf, M., Steinlein, P., et al. (2005). The profile of repeat-associated histone lysine methylation states in the mouse epigenome. *EMBO Journal*, *24*(4), 800–812.

Martienssen, R. A. (2003). Maintenance of heterochromatin by RNA interference of tandem repeats. *Nature Genetics*, *35*(3), 213–214.

Metz, A., Soret, J., Vourc'h, C., Tazi, J., & Jolly, C. (2004). A key role for stress-induced satellite III transcripts in the relocalization of splicing factors into nuclear stress granules. *Journal of Cell Science*, *117*, 4551–4558.

Mitchell, A. R. (1996). The mammalian centromere: Its molecular architecture. *Mutation Research*, *372*(2), 153–162.

Morozov, V. M., Gavrilova, E. V., Ogryzko, V. V., & Ishov, A. M. (2012). Dualistic function of Daxx at centromeric and pericentromeric heterochromatin in normal and stress conditions. *The Nucleus*, *3*(3), 276–285.

Muller, H. J. (1930). Types of visible variations induced by X-rays in Drosophila. *Journal of Genetics*, *22*, 299–335.

Nagaki, K., Cheng, Z., Ouyang, S., Talbert, P. B., Kim, M., Jones, K. M., et al. (2004). Sequencing of a rice centromere uncovers active genes. *Nature Genetics*, *36*, 138–145.

Narayan, A., Ji, W., Zhang, X. Y., Marrogi, A., Graff, J. R., Baylin, S. B., et al. (1997). Hypomethylation of pericentromeric DNA in breast adenocarcinomas. *International Journal of Cancer*, *77*(6), 833–838.

Narayan, A., Ji, W., Zhang, X. Y., Marrogi, A., Graff, J. R., Baylin, S. B., et al. (1998). Hypomethylation of pericentromeric DNA in breast adenocarcinomas. *International Journal of Cancer*, *77*(6), 833–838.

Ohno, S. (1972). So much 'junk DNA' in our genome. *Brookhaven Symposium in Biology*, *23*, 366–370.

Okada, T., Ohzeki, J., Nakano, M., Yoda, K., Brinkley, W. R., Larionov, V., et al. (2007). CENP-B controls centromere formation depending on the chromatin context. *Cell*, *131*(7), 1287–1300.

O'Neill, R. J., & Carone, D. M. (2009). The role of ncRNA in centromeres: A lesson from marsupials. *Progress in Molecular and Subcellular Biology*, *48*, 77–101.

Parris, G. E. (2010). Developmental diseases and hypothetical Master Development Program. *Medical Hypotheses*, *74*(3), 564–573.

Partridge, J. F., Borgstrom, B., & Allshire, R. C. (2000). Distinct protein interaction domains and protein spreading in a complex centromere. *Genes and Development*, *14* (7), 783–791.

Pathak, D., Srivastava, J., Premi, S., Tiwari, M., Garg, L. C., Kumar, S., et al. (2006). Transcription and evolutionary dynamics of the centromeric satellite repeat CentO in rice. *DNA and Cell Biology*, *25*(4), 206–214.

Pennisi, E. (2010). Shining a light on the genome's 'dark matter'. *Science*, *330*(6011), 1614.

Pérez, M. M., Míguez, L., Fuster, C., Miró, R., Genescà, G., & Egozcue, J. (1991). Heterochromatin decondensation in chromosomes from chorionic villus samples. *Prenatal Diagnosis*, *11*(9), 697–704.

Peters, A. H., Kubicek, S., Mechtler, K., O'Sullivan, R. J., Derijck, A. A., Perez-Burgos, L., et al. (2003). Partitioning and plasticity of repressive histone methylation states in mammalian chromatin. *Molecular Cell*, *12*(6), 1577–1589.

Pezer, Z., & Ugarkovic, D. (2012). Satellite DNA-associated siRNAs as mediators of heat shock response in insects. *RNA Biology*, *9*(5), 587–595.

Plohl, M., Luchetti, A., Mestrović, N., & Mantovani, B. (2008). Satellite DNAs between selfishness and functionality: Structure, genomics and evolution of tandem repeats in centromeric (hetero)chromatin. *Gene*, *409*(1–2), 72–82.

Podgornaia, O. I., Ostromyshenskiĭ, D. I., Kuznetsova, I. S., Matveev, I. V., & Komissarov, A. S. (2009). Heterochromatin and centromere structure paradox. *Tsitologiia*, *51*(3), 204–211.

Probst, A. V., Okamoto, I., Casanova, M., El Marjou, F., Le Baccon, P., & Almouzni, G. (2010). A strand-specific burst in transcription of pericentric satellites is required for chromocenter formation and early mouse development. *Developmental Cell*, *19*, 625–638.

Qu, G., Grundy, P. E., Narayan, A., & Ehrlich, M. (1999). Frequent hypomethylation in Wilms tumors of pericentromeric DNA in chromosomes 1 and 16. *Cancer Genetics and Cytogenetics*, *109*, 34–39.

Rand, E., & Cedar, H. (2003). Regulation of imprinting: A multi-tiered process. *Journal of Cellular Biochemistry*, *88*(2), 400–407.

Rice, J. C., Briggs, S. D., Ueberheide, B., Barber, C. M., Shabanowitz, J., Hunt, D. F., et al. (2003). Histone methyltransferases direct different degrees of methylation to define distinct chromatin domains. *Molecular Cell*, *12*(6), 1591–1598.

Ritchie, R. J., Mattei, M. G., & Lalande, M. (1998). A large polymorphic repeat in the pericentromeric region of human chromosome 15q contains three partial gene duplications. *Human Molecular Genetics*, *7*(8), 1253–1260.

Rizzi, N., Denegri, M., Chiodi, I., Corioni, M., Valgardsdottir, R., Cobianchi, F., et al. (2004). Transcriptional activation of a constitutive heterochromatic domain of the human genome in response to heat shock. *Molecular Biology of the Cell*, 15(2), 543–551.

Rouleux-Bonnin, F., Bigot, S., & Bigot, Y. (2004). Structural and transcriptional features of *Bombus terrestris* satellite DNA and their potential involvement in the differentiation process. *Genome*, 47, 877–888.

Rouleux-Bonnin, F., Renault, S., Bigot, Y., & Periquet, G. (1996). Transcription of four satellite DNA subfamilies in *Diprion pini* (Hymenoptera, Symphyta, Diprionidae). *European Journal of Biochemistry*, 238, 752–759.

Ruault, M., van der Bruggen, P., Brun, M. E., Boyle, S., Roizès, G., & De Sario, A. (2002). New BAGE (B melanoma antigen) genes mapping to the juxtacentromeric regions of human chromosomes 13 and 21 have a cancer/testis expression profile. *European Journal of Human Genetics*, 10(12), 833–840.

Rudert, F., Bronner, S., Garnier, J. M., & Dollé, P. (1995). Transcripts from opposite strands of gamma satellite DNA are differentially expressed during mouse development. *Mammalian Genome*, 6(2), 76–83.

Rudert, F., & Gronemeyer, H. (1993). Retinoic acid-response elements with a highly repetitive structure isolated by immuno-selection from genomic DNA. *The Journal of Steroid Biochemistry and Molecular Biology*, 46(2), 121–133.

Saffery, R., Sumer, H., Hassan, S., Wong, L. H., Craig, J. M., Todokoro, K., et al. (2003). Transcription within a functional human centromere. *Molecular Cell*, 12(2), 509–516.

Sandqvist, A., Björk, J. K., Akerfelt, M., Chitikova, Z., Grichine, A., Vourc'h, C., et al. (2009). Heterotrimerization of heat-shock factors 1 and 2 provides a transcriptional switch in response to distinct stimuli. *Molecular Biology of the Cell*, 20(5), 1340–1347.

Santenard, A., Ziegler-Birling, C., Koch, M., Tora, L., Bannister, A. J., & Torres-Padilla, M. E. (2010). Heterochromatin formation in the mouse embryo requires critical residues of the histone variant H3.3. *Nature Cell Biology*, 12(9), 853–862.

Sarge, K. D., Murphy, S. P., & Morimoto, R. I. (1993). Activation of heat shock gene transcription by heat shock factor 1 involves oligomerization, acquisition of DNA-binding activity, and nuclear localization and can occur in the absence of stress. *Molecular and Cellular Biology*, 13(3), 1392–1407.

Schnedl, W., Mikelsaar, A. V., Breitenbach, M., & Dann, O. (1977). DIPI and DAPI: Fluorescence banding with only negliglible fading. *Human Genetics*, 36(2), 167–172.

Schueler, M. G., Higgins, A. W., Rudd, M. K., Gustashaw, K., & Willard, H. F. (2001). Genomic and genetic definition of a functional human centromere. *Science*, 294(5540), 109–115.

Schueler, M. G., & Sullivan, B. A. (2006). Structural and functional dynamics of human centromeric chromatin. *Annual Review of Genomics and Human Genetics*, 7, 301–313.

Seidman, M. M., & Cole, R. D. (1977). Chromatin fractionation related to cell type and chromosome condensation but perhaps not to transcriptional activity. *Journal of Biological Chemistry*, 252(8), 2630–2639.

Sengupta, S., Parihar, R., & Ganesh, S. (2009). Satellite III non-coding RNAs show distinct and stress-specific patterns of induction. *Biochemical and Biophysical Research Communications*, 382(1), 102–107.

She, X., Horvath, J. E., Jiang, Z., Liu, G., Furey, T. S., Christ, L., et al. (2004). The structure and evolution of centromeric transition regions within the human genome. *Nature*, 430 (7002), 857–864.

Shelby, R. D., Hahn, K. M., & Sullivan, K. F. (1996). Dynamic elastic behavior of alpha-satellite DNA domains visualized in situ in living human cells. *The Journal of Cell Biology*, 135(3), 545–557.

Shestakova, E. A., Mansuroglu, Z., Mokrani, H., Ghinea, N., & Bonnefoy, E. (2004). Transcription factor YY1 associates with pericentromeric gamma-satellite DNA in cycling but not in quiescent (G0) cells. *Nucleic Acids Research*, 32(14), 4390–4399.

Shumaker, D. K., Dechat, T., Kohlmaier, A., Adam, S. A., Bozovsky, M. R., Erdos, M. R., et al. (2006). Mutant nuclear lamin A leads to progressive alterations of epigenetic control in premature aging. *Proceedings of the National Academy of Sciences of the United States of America, 103*(23), 8703–8708.

Skibbens, R. V., Skeen, V. P., & Salmon, E. D. (1993). Directional instability of kinetochore motility during chromosome congression and segregation in mitotic newt lung cells: A push-pull mechanism. *The Journal of Cell Biology, 122*(4), 859–875.

Smith, C. D., Shu, S., Mungall, C. J., & Karpen, G. H. (2007). The Release 5.1 annotation of *Drosophila melanogaster* heterochromatin. *Science, 316*(5831), 1586–1591.

Sreedhar, A. S., & Csermely, P. (2004). Heat shock proteins in the regulation of apoptosis: New strategies in tumor therapy: A comprehensive review. *Pharmacology and Therapeutics, 101*(3), 227–257.

Sturtevant, A. H. (1965). *A history of genetics*. New York: Harper and Row.

Sugimoto, K., Kuriyama, K., Shibata, A., & Himeno, M. (1997). Characterization of internal DNA-binding and C-terminal dimerization domains of human centromere/kinetochore autoantigen CENP-C in vitro: Role of DNA-binding and self-associating activities in kinetochore organization. *Chromosome Research, 5*(2), 132–141.

Sugimura, K., Fukushima, Y., Ishida, M., Ito, S., Nakamura, M., Mori, Y., et al. (2010). Cell cycle-dependent accumulation of histone H3.3 and euchromatic histone modifications in pericentromeric heterochromatin in response to a decrease in DNA methylation levels. *Experimental Cell Research, 316*(17), 2731–2746.

Sullivan, K. F., Hechenberger, M., & Masri, K. (1994). Human CENP-A contains a histone H3 related histone fold domain that is required for targeting to the centromere. *The Journal of Cell Biology, 127*(3), 581–592.

Sullivan, B. A., & Karpen, G. H. (2004). Centromeric chromatin exhibits a histone modification pattern that is distinct from both euchromatin and heterochromatin. *Nature Structural and Molecular Biology, 11*(11), 1076–1083.

Suzuki, T., Fujii, M., & Ayusawa, D. (2002). Demethylation of classical satellite 2 and 3 DNA with chromosomal instability in senescent human fibroblasts. *Experimental Gerontology, 37*(8–9), 1005–1014.

Tessadori, F., Chupeau, M. C., Chupeau, Y., Knip, M., Germann, S., van Driel, R., et al. (2007). Large-scale dissociation and sequential reassembly of pericentric heterochromatin in dedifferentiated Arabidopsis cells. *Journal of Cell Science, 120*(Pt 7), 1200–1208.

Tessadori, F., Schulkes, R. K., van Driel, R., & Fransz, P. (2007). Light-regulated large-scale reorganization of chromatin during the floral transition in Arabidopsis. *The Plant Journal, 50*(5), 848–857.

Ting, D. T., Lipson, D., Paul, S., Brannigan, B. W., Akhavanfard, S., Coffman, E. J., et al. (2011). Aberrant overexpression of satellite repeats in pancreatic and other epithelial cancers. *Science, 331*(6017), 593–596.

Trapitz, P., Wlaschek, M., & Bünemann, H. (1988). Structure and function of Y chromosomal DNA. II. Analysis of lampbrush loop associated transcripts in nuclei of primary spermatocytes of *Drosophila hydei* by in situ hybridization using asymmetric RNA probes of four different families of repetitive DNA. *Chromosoma, 96*(2), 159–170.

Ugarkovic, D. (2005). Functional elements residing within satellite DNAs. *EMBO Reports, 6* (11), 1035–1039.

Valgardsdottir, R., Chiodi, I., Giordano, M., Cobianchi, F., Riva, S., & Biamonti, G. (2005). Structural and functional characterization of noncoding repetitive RNAs transcribed in stressed human cells. *Molecular Biology of the Cell, 16*(6), 2597–2604.

Valgardsdottir, R., Chiodi, I., Giordano, M., Rossi, A., Bazzini, S., Ghigna, C., et al. (2008). Transcription of Satellite III non-coding RNAs is a general stress response in human cells. *Nucleic Acids Research, 36*(2), 423–434.

Vansant, G., & Reynolds, W. F. (1995). The consensus sequence of a major Alu subfamily contains a functional retinoic acid response element. *Proceedings of the National Academy of Sciences of the United States of America, 92*(18), 8229–8233.

Varadaraj, K., & Skinner, D. M. (1994). Cytoplasmic localization of transcripts of a complex G+C-rich crab satellite DNA. *Chromosoma, 103*, 423–431.

Varley, J. M., Macgregor, H. C., Nardi, I., Andrews, C., & Erba, H. P. (1980). Cytological evidence of transcription of highly repeated DNA sequences during the lampbrush stage in *Triturus cristatus* carnifex. *Chromosoma, 80*(3), 289–307.

Vissel, B., & Choo, K. H. (1989). Mouse major satellite is highly conserved and organized into extremely long tandem arrays: Implications for recombination between non-homologous chromosome 6 and its response to interferon in interphase nuclei. *Journal of Cell Science, 113*, 1565–1576.

Vissel, B., Nagy, A., & Choo, K. H. (1992). A satellite III sequence shared by human chromosomes 13, 14, and 21 that is contiguous with alpha satellite DNA. *Cytogenetics and Cell Genetics, 61*(2), 81–86.

Vourc'h, C., & Biamonti, G. (2011). Transcription of satellite DNAs in mammals. *Progress in Molecular and Subcellular Biology, 51*, 95–118.

Wilson, A. S., Power, B. E., & Molloy, P. L. (2007). DNA hypomethylation and human diseases. *Biochimica et Biophysica Acta, 1775*(1), 138–162.

Wong, L. H., Brettingham-Moore, K. H., Chan, L., Quach, J. M., Anderson, M. A., Northrop, E. L., et al. (2007). Centromere RNA is a key component for the assembly of nucleoproteins at the nucleolus and centromere. *Genome Research, 17*(8), 1146–1160.

Woo, S. K., Lee, S. D., Na, K. Y., Park, W. K., & Kwon, H. M. (2002). TonEBP/NFAT5 stimulates transcription of HSP70 in response to hypertonicity. *Molecular and Cellular Biology, 22*(16), 5753–5760.

Yan, H., & Jiang, J. (2007). Rice as a model for centromere and heterochromatin research. *Chromosome Research, 15*, 77–84.

Yan, H., Jin, W., Nagaki, K., Tian, S., Ouyang, S., Buell, C. R., et al. (2005). Transcription and histone modifications in the recombination-free region spanning a rice centromere. *The Plant Cell, 17*, 3227–3238.

Yasuhara, J. C., & Wakimoto, B. T. (2006). Oxymoron no more: The expanding world of heterochromatic genes. *Trends in Genetics, 22*(6), 330–338.

Zhao, J., Morozova, N., Williams, L., Libs, L., Avivi, Y., & Grafi, G. (2001). Two phases of chromatin decondensation during dedifferentiation of plant cells: Distinction between competence for cell fate switch and a commitment for S phase. *Journal of Biological Chemistry, 276*(25), 22772–22778.

Zhu, Q., Pao, G. M., Huynh, A. M., Suh, H., Tonnu, N., Nederlof, P. M., et al. (2011). BRCA1 tumour suppression occurs via heterochromatin-mediated silencing. *Nature, 477*(7363), 179–184.

CHAPTER THREE

Zinc Finger Proteins and the 3D Organization of Chromosomes

Christoph J. Feinauer*, Andreas Hofmann*, Sebastian Goldt†, Lei Liu*, Gabriell Máté*, Dieter W. Heermann*,‡,§,1

*Institute for Theoretical Physics, Heidelberg University, Philosophenweg, Heidelberg, Germany
†Fitzwilliam College, Cambridge University, Cambridge, United Kingdom
‡The Jackson Laboratory, Bar Harbor, Maine, USA
§Shanghai Institute of Biological Sciences (SIBS), Chinese Academy of Sciences (CAS), Shanghai, PR China
1Corresponding author: e-mail address: heermann@tphys.uni-heidelberg.de

Contents

Abstract

Zinc finger domains are one of the most common structural motifs in eukaryotic cells, which employ the motif in some of their most important proteins (including TFIIIA, CTCF, and ZiF268). These DNA binding proteins contain up to 37 zinc finger domains connected by flexible linker regions. They have been shown to be important organizers of the 3D structure of chromosomes and as such are called the master weaver of the genome.

Using NMR and numerical simulations, much progress has been made during the past few decades in understanding their various functions and their ways of binding to the DNA, but a large knowledge gap remains to be filled. One problem of the hitherto existing theoretical models of zinc finger protein DNA binding in this context is that they are aimed at describing specific binding. Furthermore, they exclusively focus on the microscopic details or approach the problem without considering such details at all. We present the *Flexible Linker Model*, which aims explicitly at describing nonspecific binding. It takes into account the most important effects of flexible linkers and allows a qualitative investigation of the effects of these linkers on the nonspecific binding affinity of zinc finger proteins to DNA.

Our results indicate that the binding affinity is increased by the flexible linkers by several orders of magnitude. Moreover, they show that the binding map for proteins with more than one domain presents interesting structures, which have been neither observed nor described before, and can be interpreted to fit very well with existing theories of facilitated target location. The effect of the increased binding affinity is also in agreement with recent experiments that until now have lacked an explanation.

We further explore the class of proteins with flexible linkers, which are unstructured until they bind. We have developed a methodology to characterize these flexible proteins. Employing the concept of barcodes, we propose a measure to compare such flexible proteins in terms of a similarity measure. This measure is validated by a comparison between a geometric similarity measure and the topological similarity measure that takes geometry as well as topology into account.

1. THE 3D ORGANIZATION OF THE GENOME AND LOOPS

The large-scale organizations of the genome have puzzled researchers for a long time. Some of the most important questions are, for example: Why do chromosomes occupy discrete territories in the nucleus (Cremer & Cremer, 2010), that is, why do chromosomes not mix in interphase? Why do chromosomes internally segregate into structural compartments or domains (Jerabek & Heermann, 2012; Verschure, van Der Kraan, Manders, & van Driel, 1999)? Some key experiments have pointed to the fact that loops may play a central role in this. Thus, the interaction of loci genomically very far apart along a chromosome impact on the overall organization and structures of the chromosomes as well as the nucleus.

Early fluorescence *in situ* hybridization experiments probed chromatin–chromatin interactions (Mateos-Langerak et al., 2009; Sexton, Bantignies, & Cavalli, 2009; Simonis, Kooren, & de Laat, 2007) and showed that genomically distant parts of the chromosome colocalized or were at least in physical proximity. If two genomically distant parts of the chromosome come together, then this establishes a physical loop with the loop sizes found ranging from a few kilobytes up to tens of megabytes (Lieberman–Aiden et al., 2009; Sexton et al., 2009; Simonis et al., 2006). Hence, chromosomes have to be considered not as linear chains but rather as looped structures (Bohn & Heermann, 2010; Heermann, 2011). That loops are indeed ubiquitous has been established by biochemical analyses such as chromosome conformation capture techniques (Dekker, 2003; de Laat & Grosveld, 2003; Lieberman–Aiden et al., 2009).

This has severe consequences on the effects that chromosomes have on each other as well as on the internal organization (see Fig. 3.1). The simple topological alteration from a situation of open ends as opposed to a closed situation changes the entropy significantly. Two linear chains mix (assuming that they are basically the same in a chemical sense) while there is an entropic repulsion between the two loops (Fritsche et al., 2012; Heermann, 2012; Zhang & Heermann, 2011). This is the key to understanding the organization: Since loops entropically repel each other, two looped structures repel each other. Thus the chromosomes do not mix and the internal segregation into the differently looped structures takes place.

There is no fixed loop structure. It controls and is controlled by genome activities: transcription, epigenetic repression, and replication. The loop structure also varies in time and has a considerable cell-to-cell variation.

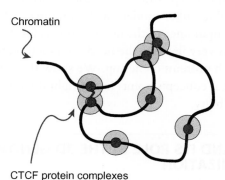

Chromatin

CTCF protein complexes

Figure 3.1 The organization of the genome is strongly influenced by loops. These loops are to a large extent based on CTCF (CCCTC-binding factor (zinc finger protein)) complexes. Chromosome territories as well as the internal structure of the chromosomes depend on the binding of CTCF and the complexes that CTCF forms with other proteins like cohesin. However, little is known about how CTCF influences the genome organization and the properties that singled out CTCF as an organizer. (For color version of this figure, the reader is referred to the online version of this chapter.)

Unknown though to a larger extent is the mechanism for looping. Much attention has been given to the role of CTCF (CCCTC-binding factor (zinc finger protein)) in chromatin looping (Filippova, 2008; Wendt & Peters, 2009; Zlatanova & Caiafa, 2009). CTCF has been shown to correlate highly with the looping site exposed in the Hi-C chromosome conformation capture experiments (Botta, Haider, Leung, Lio, & Mozziconacci, 2010). Regulation of gene expression has been shown. Also, the existence of processes such as bringing together or isolating regulatory sequences (including promoters and enhancers) from each other has been verified. Specific attention has been given to the developmentally controlled beta-globin locus and the imprinted H19–Igf2 locus (Kurukuti et al., 2006; Splinter et al., 2006). These studies all point to the important role that CTCF plays in the organization and maintenance of the 3D structure of the genome.

Another key player is cohesin. CTCF looping requires cohesin (Hadjur et al., 2009; Mishiro et al., 2009; Rubio et al., 2008). And, there are also reports of non CTCF-induced looping, for example, binding to the nuclear lamina (Guelen et al., 2008). Here, however, we focus on properties of CTCF and its binding models to the DNA.

In this chapter, we investigate properties that single out CTCF as an organizer. In particular, we first look at the linkers of CTCF. We show that the linkers bestow the protein with a flexibility which is consistent with that of a random or self-avoiding walk. Thus CTCF can be considered a flexible polymer that can take on a plethora of possible conformations. Based on this, we establish a flexible linker model for the binding to DNA. This model clearly shows the importance of the flexible linkers for the binding patterns. Since the protein is very flexible, the need for a method to investigate how ligands bind to such proteins comes up. We show that a similarity measure based on topological concepts points in the right direction of how one can deal with the flexibility.

2. CTCF AND ITS ROLE IN THE 3D GENOME ORGANIZATION

The zinc finger domain is one of the most common structural motifs in the proteins of eukaryotes (Pabo, Peisach, & Grant, 2001) and some of the cell's key proteins are zinc finger proteins like ZiF268 (Egr-1) (Zandarashvili et al., 2012), transcription factor IIIA (Pabo et al., 2001), and CTCF, the putative *Master Weaver of the Genome* (Phillips & Corces, 2009). These proteins are often array-like structures of several zinc finger domains connected by flexible linker regions. The number of domains ranges from 3 in ZiF268 (Pavletich & Pabo, 1991) to 11 for CTCF (Phillips & Corces, 2009), to 37 in the *Xenopus* developmental protein Xfin (De Lucchini & Cardellini, 2000) (see Fig. 3.2).

Besides their importance for the cell, the amount of attention given to zinc finger proteins by researchers in the past 20 years has been augmented by another point: The apparently simple modular way cellular evolution joins several zinc finger domains together to create proteins with specific functions and specific binding targets on the DNA. Expectations were that

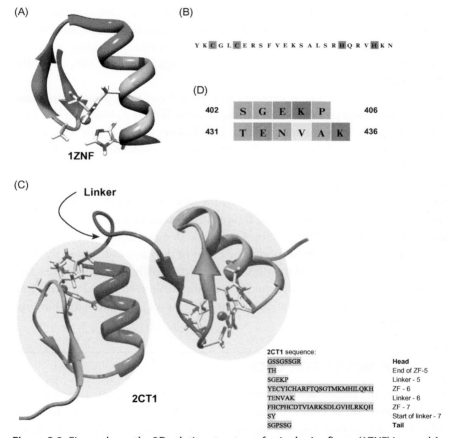

Figure 3.2 Figure shows the 3D solution structure of a single zinc finger (1ZNF) in panel A. Panel B shows the corresponding sequence. Panel C shows 2CT1 with zinc finger domains and the corresponding linkers. The linkers are flexible polymers. The head and tail (see panel D) do not belong to the protein. (For color version of this figure, the reader is referred to the online version of this chapter.)

by mimicking this strategy, a human designer could artificially create modified versions of known or even completely new proteins with predictable characteristics (Kim & Pabo, 1998; Pabo et al., 2001; Rebar & Pabo, 1994). The applications in research and drug design seemed endless. Although the truth turned out to be more complex and zinc finger recognition patterns were less modular than initially hoped (Paillard, Deremble, & Lavery, 2004), the field is still one of active and promising research (Sander, Zaback, Keith Joung, Voytas, & Dobbs, 2007).

Despite the importance of zinc finger proteins, the binding structure of the specifically bound complex and the exact origin of the sequence specificity are only beginning to be elucidated (Paillard et al., 2004). Moreover, for nonspecific binding, very basic topics such as the number of zinc finger domains a zinc finger protein actually uses in nonspecific binding with the DNA were neither known nor under investigation until very recently (Zandarashvili et al., 2012).

An aspect of zinc finger proteins that turned out to be more important than might appear at a first glance is the flexible linkers connecting the zinc finger domains. Experiments indicate that linkers play a major role in the sequence specificity (Ryan & Darby, 1998) and are a determinant of the general affinity to DNA. In fact, compromising the linker's flexibility, for example, by phosphorylation, reduces the binding affinity to DNA by orders of magnitude (Jantz & Berg, 2004). Also, the structuring of the linker region seems to be a key mechanism of fixing the protein in a specifically bound state (Hyre & Klevit, 1998; Laity, Dyson, & Wright, 2000).

There are approaches to model the effect of domains connected by flexible linkers on DNA binding affinity, but they usually use a rather macroscopic description directed at specific binding limited to two domains, not taking into account the statistical implications of multidomain binding (Zhou, 2001).

3. A MODEL FOR CTCF

The zinc finger domain comprises, as its name suggests, a central zinc ion. This zinc ion does not interact with the DNA but stabilizes the protein, which consists of two β-sheets and one α-helix. There are several versions of this domain and the exact definition of the term is under debate (Pabo et al., 2001). The most common version, though, is the "classical zinc finger" with a distinct Cys_2His_2 structure, shown in Fig. 3.2. Zinc finger domains bind to the major groove in the DNA, where specific amino acid residues contact bases by hydrogen bonding. The whole domain comprises about 30 amino acids. The interaction with DNA is typically weak, and this might be one reason why a zinc finger protein usually has several domains (Nelson & Cox, 2004).

Pabo et al. (2001) mention that zinc finger domains (specifically the Cys_2His_2 domain) may interact with RNA as well and binding to other proteins has also been observed. Indeed, the complexation of the 11-zinc finger protein CTCF with other proteins, for example, cohesin, might be a central theme in cellular organization (Feeney & Verma-Gaur, 2012). It is thus of importance to understand on the one hand the binding pattern and affinity to DNA and on the other, the affinity to other proteins. As at least one of the binding partners (CTCF) is a flexible protein, the typical strategy of ligation prediction does not apply. In a later section, we thus formulate a strategy to attack this problem.

Zinc finger proteins generally comprise more than one zinc finger domain that binds independently to DNA. The 11-zinc finger protein CTCF, for example, binds only with a certain subset of its fingers, but this subset changes, depending on the context and probably function (Ohlsson, Renkawitz, & Lobanenkov, 2001). This could be another strength of zinc finger proteins, enabling the cell to use one single zinc finger protein for different tasks encoded in the subset of fingers binding to the DNA (note that this is unproven so far).

Whether the basic length for a zinc finger DNA contact is 3 or 4 bp is under debate and differently depicted in different publications (Pabo et al., 2001; Paillard et al., 2004).

The zinc finger domains are connected by linker regions. These linker regions seem to be important for the binding affinity and altering their structural characteristics strongly decreases the binding affinity of the whole protein (Elrod-Erickson, Rould, Nekludova, & Pabo, 1996; Laity et al., 2000).

One of the major points of interest in zinc finger proteins stemmed from their apparent modularity. Depending on which zinc finger domain variant is found in the zinc finger protein, the sequences to which the protein binds preferentially changes. For a long time, it was assumed that the individual fingers of a protein act independently in this process, and the binding sequence for the whole protein is simply the linear concatenation of the binding sequences of the individual fingers. This turned out to be partly wrong. First of all, it is not really clear how the binding sequence of even small zinc finger proteins follows from their amino acid sequence and furthermore, upon specific binding, the whole protein undergoes conformational changes (Hyre & Klevit, 1998). This makes it probable that the individual zinc finger domains interact and the sequence preference of the concatenated zinc finger domains is not just an independent addition of the sequence preferences of the individual domains.

Notwithstanding these setbacks, the design and selection of zinc finger proteins is a promising field of research (Pabo et al., 2001). The ability to easily construct proteins that bind specific sequences would be of enormous use in science and in medicine. Especially, if the way in which the nonbound fingers interact with other proteins is elucidated, it could create a technique to put a specific (maybe novel) function at a specific sequence of the DNA.

3.1. CTCF as an unstructured protein with flexible linkers

A feature that is very important for CTCF is the flexibility of the linkers. This allows the protein to be quasi unstructured. This flexibility has an enormous impact on the conformational properties. In Fig. 3.3, we show two such conformations. The question that arises in this context is whether CTCF can be considered to be in the same class as the selfavoiding random walks (de Gennes, 1979) (i.e., polymers with excluded volume). Thus CTCF would be a polymer made up of monomers connected by bonds. In this class are the polymers with conformational properties that are almost those of a random walk except that those conformations which would violate the excluded volume requirement (two monomers of a polymer cannot sit on top of each other) are discarded.

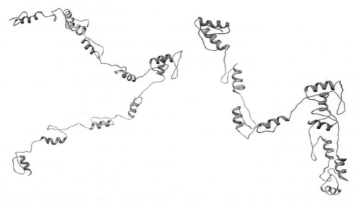

Figure 3.3 Two conformations of the CTCF protein. The conformations are determined by an effective monomer size. (For color version of this figure, the reader is referred to the online version of this chapter.)

3.2. Conformational properties of CTCF

To investigate the conformational properties of CTCF, we use a Monte Carlo method (Heermann & Binder, 2010) to sample the possible conformations that CTCF admits. For this, we randomly pick pivot points along the backbone of CTCF. Then a random rotation around the pivot point is made (Madras & Sokal, 1988). This conformation is accepted if no excluded volume violation is detected. In this case, an energy minimization of the conformation is performed. After the energy minimum is achieved, the conformation is accepted as a new conformation, and the pivot process is repeated until a minimum number of conformations are sampled.

Since CTCF is rather short, the difference between a random and a selfavoiding walk is rather small. Thus, we have chosen to analyze the conformations only in terms of a random walk behavior. Let N denote the number of zinc finger monomers joined by linkers and b the effective monomer size. Then for a random walk the end–to–end distance R that one finds can be calculated as

$$\langle R^2 \rangle = b^2 N \tag{3.1}$$

Here $\langle \cdot \rangle$ denotes the average over all possible conformations. The distribution of the end–to–end distances is given by

$$P(R,N) = 2\pi R^2 \left(\frac{3}{2\pi N b^2} \right) \exp - \frac{3R^2}{2Nb^2} \tag{3.2}$$

In Fig. 3.4 are shown the distributions for three cases: 3, 7, and 11 zinc fingers. The expected distributions of the three cases fit the sampled conformations well. From the conformations were calculated the average end–to–end distances as shown in panel B. From this, we can extract the average monomer size $b = 27.16$ Å, which is slightly larger than a single zinc finger.

Also, the result shown in panel C where the effective monomer size is extracted from the fitted distribution parameters is consistent with the theoretical expectation and gives a single value. Thus, CTCF can be considered a polymer composed of monomers with flexible linkers, where the effective monomer size is slightly larger than a single zinc finger.

4. A MODEL FOR CTCF BINDING

We now analyze the nonspecific binding behavior of multidomain zinc finger proteins with flexible linkers in the presence of a sequence-dependent potential. By sequence dependence, we mean that the binding

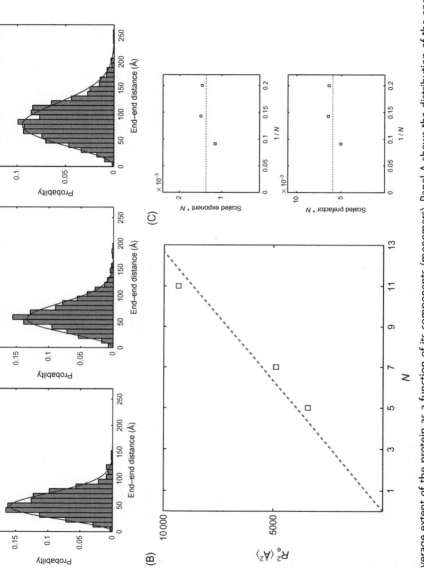

Figure 3.4 Average extent of the protein as a function of its components (monomers). Panel A shows the distribution of the end-to-end distance for 5, 7, and 11 zinc fingers. The red curve shows the fit with the theoretic prediction for a random walk. The fit with a self-avoiding walk on this scale is almost identical. Panel B shows the average end-to-end squared as a function for the number of zinc fingers. The straight line shows the fit with the random walk model. Again, on this scale, the difference with the self-avoiding walk is negligible. Panel C shows the parameter extracted from the fits shown in Panel A. That the three fits give the same value for the parameter shows the consistency yielding the effective monomer size of $b \approx 27.16$ Å slightly larger than a single zinc finger. (For interpretation of the references to color in this figure

energy of a DNA–protein contact depends on the contact's position on the DNA. We note here directly that the scope is to look at areas with increased binding energy with respect to a background rather than a real nucleotide base sequence. To this end, we develop the *flexible linker model*, which takes the most important aspects of flexible linkers into account. The model is developed in a general framework and can be used for every zinc finger protein. It is validated for the three-zinc finger protein ZiF268 because for this protein data are available. The model and the qualitative results should hold nonetheless for CTCF as well.

We show that the sequence-dependent potential not only increases the binding affinity by orders of magnitude, but also changes the way a zinc finger protein binds. Namely, the zinc finger protein preferably binds with different numbers of domains depending on the potential. Hence, already in the (simple) flexible linker model, the binding energy affects the structure of a nonspecific binding complex. Taking into account the simple assumptions, we make this is a remarkably strong statement.

Independently, this observation can be interpreted in connection with recent theories: In the wake of the idea of "Search and Recognition States" (Slutsky & Mirny, 2004), structure changing effects of position-dependent potentials were conjectured, but no microscopic interpretation was given. For zinc finger proteins, our model predicts such effects independently of these theories, but that the effects we observe *are* the effects which were postulated in them is an obvious interpretation.

In this context, more bound domains correspond to a slower diffusion but an increased propensity to recognize a correct binding site because they resemble more the specifically bound complex (with all domains bound, e.g., for ZiF268). Less bound domains correspond to an increased diffusion coefficient but a smaller propensity to recognize a correct state. The probability to find the protein in one or the other state is directly related to the sequence-dependent potential.

The outcome of Zandarashvili's recent experiment (Zandarashvili et al., 2012) serves as an argument in favor of this switch process. NMR experiments as well as MD simulations show that the zinc finger protein ZiF268 (also called Egr-1) binds with only a subset of its three fingers to DNA when it binds nonspecifically. This is surprising for Zandarashvili because the specific complex is known to always involve all three zinc fingers. In the light of the insights following from the flexible linker model, a lower number is to be expected: The interplay of the chemical potential, the binding energy, and the entropy cause the most likely state for small binding energies to be one of a partially bound zinc finger protein.

4.1. Formulation of the model

Before we describe the model it is necessary to express explicitly what we want to model: The nonspecific binding of modular zinc finger proteins at thermal equilibrium to unfolded, nucleosome-free DNA. This statement already conveys several assumptions about the underlying physical system and in order to elucidate what these assumptions mean and why they are necessary, we will describe the three major concepts here in detail.

• *Nonspecificity*: One of the key concepts of the model is that the qualities of the single DNA–protein contacts are similar. We therefore assume the protein and the DNA to have the same chemical and topological properties at every binding position. Also, the binding energy the single units contribute is assumed to be additive (this will be elaborated under *modularity*). Upon binding to its specific site, though, this assumption is certainly violated for zinc finger proteins. A protein generally undergoes conformational changes (along with the DNA) upon binding to specific target sites. Specifically for zinc finger proteins, Hyre and Klevit (1998) show this for the yeast protein ADR1. Upon binding to its specific site, although the zinc fingers themselves do not change their secondary structure substantially, this protein shows a strong structuring of the linkers and terminal regions that enhance the effective binding energy strongly. To exclude such effects, we define the model to simulate only nonspecific binding.

• *Modularity*: The topic of modular binding of zinc finger proteins arises twice: First for the single residues of one zinc finger domain and then for the separate domains of the whole zinc finger protein. The basic meaning of *modularity* for the individual zinc finger is that the binding energy of its DNA contacts is additive. For the individual zinc finger contacting 3–4 bp on the DNA, this means that we can look at the single contact energies and add them up to get the binding energy for the complete protein. For the whole zinc finger protein, the modular binding of its individual domains is important: The binding energy of the domains is additive as the binding energy of the single residues of one zinc finger domain is additive. Also, the binding energy depends only on how many domains are bound, not on *which* domains are bound.

• *Thermal Equilibrium*: The time for which we look at the system should be long enough for the ergodic hypothesis to hold; else the statistical analysis carried out for the model would not be justified. Although this matter lies at the heart of many biophysical problems, no satisfying answer to this question has been given so far. There is, though, a strong indication in favor of thermal equilibrium models that Segal mentions in Segal and Widom (2009) as a justification for this assumption: They work and give correct predictions for real experiments.

The principle entities of the model are the proteins and the DNA polymer, which are both represented by linear 1D chains of finite size (see Fig. 3.5).

The DNA polymer is defined as a one-dimensional array of *units*, and its state as the sequence of states s_i of all its units numbered by $i = 1, 2, \ldots, N$. The state of a unit (often referred to as 1 bp for one base pair) is basically defined by whether it is bound or free, and if bound, also by what part of which protein it is bound. The s_i might then be unique numerical identifiers for these binding states (e.g., $s_1 = 1$ for *the first unit of the DNA is bound by the first position of a protein* if there is only one protein type).

A protein is represented by its *DNA binding interface*, also a one-dimensional array of units. This is often directly referred to as the protein because beside its binding interface, no part of the protein is present in the model. Like the DNA, it is defined by the sequence of its constituent units. One unit (often also referred to as 1 bp) of the DNA binding interface can bind to one unit of the DNA.

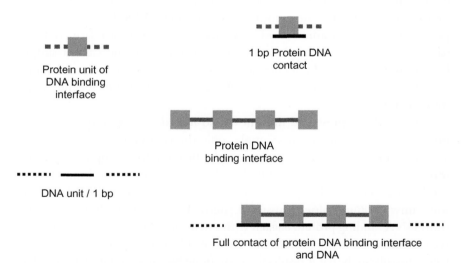

Protein unit of
DNA binding
interface

1 bp Protein DNA
contact

Protein DNA
binding interface

DNA unit / 1 bp

Full contact of protein DNA binding interface
and DNA

Figure 3.5 Basic entities and concepts of the model and their graphical representations. (For color version of this figure, the reader is referred to the online version of this chapter.)

The protein unit is not necessarily equal to one special amino acid, although such a direct correspondence can often be given by experiments (e.g., for zinc fingers by Pabo et al., 2001). The definition of the protein in the model is operational. For simplicity, we choose to define the number of units of a protein as equal to the number of DNA units that bind to it. In this way we count, for example, a whole group of amino acids as one protein unit if only one of the amino acids can make a contact. For zinc finger proteins, this means that one zinc finger domain corresponds to three or four protein units.

Usually, there is only one DNA polymer in the system (system here means DNA plus proteins bound to it) while the possible number of proteins is infinite. The actual number on the DNA is limited, though, by a finite chemical potential and the requirement that every unit of the DNA can be bound only by one protein unit.

There can be several protein types in the system, which may, for example, differ by their lengths, their binding energy, or their chemical potential. When a protein binds in the model, it does so in a *directed* and *ordered* way with respect to its units. The first assumption means that a protein binding to a DNA sequence in $5' \rightarrow 3'$ direction on a strand cannot bind the same position on the same strand in $3' \rightarrow 5'$ direction. The second assumption then defines that the order of the protein units relative to each other is fixed. This stems from the fact that the units that comprise a DNA binding interface of a protein represent an amino acid sequence embedded in a relatively fixed structure with a hierarchical organization of topology and cannot simply be permuted. The zinc finger domain, for example, will not twist its secondary structure and make contacts in a different order than at its specific binding domain.

The model as presented until now also includes the possibility to bind only pieces of a protein, for example, only the two units in the middle of a 4-bp spanning protein or maybe just the first and last 50 bp of a nucleosome, such that the 47 bp of DNA still free between those binding sites can be bound by another protein. This is a real effect called partial nucleosome unwrapping (Poirier, Bussiek, Langowski, & Widom, 2008), but for a short DNA binding protein *domain* with a dedicated DNA binding interface, we can safely assume that this is not the case. The topology of such an interface is usually such that it results in all-or-none binding, and it seems hard to prevent a nucleotide–amino acid contact if both are already fixed in the correct position (see e.g., Pabo on the zinc finger domain in the major groove; Pabo et al., 2001).

Note that this holds only for single protein domains. For several protein domains with flexible linkers, it is *known* that partial binding of the protein (but full binding of the single domains) occurs. Zandarashvili et al. (2012) show NMR and MD evidence for this which will be treated in greater detail later. Partial binding is introduced in this sense into the model.

We also choose not to allow binding over the end of the chain. This unimportant looking assumption leads to effects at the end of the chain that can explain nucleosome positioning by insulators or other "road blocks" on the DNA as (Kornberg and Stryer (1988) deduced analytically.

We, however, are not directly interested in end effects and only look at the middle regions of long chains. There we can expect those end effects to have a negligible influence. At the same time, the observation of those effects when looking at the whole chain can serve as a confirmation that the implementation is sound.

Figure 3.5 shows the principle entities and concepts of the model and the graphical representation used throughout this paper (the mathematical formulation is deferred to that in Appendix).

The 1D-lattice model does not necessarily prohibit the inclusion of effects of 3D structure, which are at least in part defined by structural proteins bound to the DNA. Intrinsic bends of the DNA could, for example, be introduced as potentials affecting the binding energy of other factors. Binding proteins could themselves create bends and therefore change the position-dependent potential in the vicinity of their binding site; as another example, looping proteins could be simulated by a protein that can bind to two distant units of the DNA. It could then introduce bending of the intermediate region or function as an insulator at the binding sites (as CTCF does; Phillips & Corces, 2009).

On the other hand, the effects of those topologically acting proteins would need to be artificially introduced into the system and not follow from it. The direct simulation would therefore be preferable if the simulation of the 3D structure itself was the scope (e.g., by Monte Carlo lattice models (Bohn & Heermann, 2010) or MD simulations (de Pablo, 2011)).

As an example, the effect of a bound CTCF molecule (presumably building a loop) could be presented at least partly in the way described earlier, but the energy of building the loop is heavily dependent on the change of entropy of the chain by the loop. To implement this, one would have to introduce some measure of the change in free energy apart from the binding energy and take a random walk or a more elaborate model as a reference point. And even then one would have neglected the fact that this change in energy depends on the state of the whole chain, not only on the looped part. In MC and MD simulations of DNA polymers, such effects appear in the simulation directly and they can even be measured.

Because of these difficulties and because the simulation of 3D structure is not our scope, we assume that the topological effects of binding proteins do not influence the binding of other proteins. We keep in mind, though, that this is probably the single most simplifying assumption of our model. Still, we note that those factors will become more important at longer chain length while we explicitly model binding to short, free DNA pieces.

4.2. Flexible linkers and domains

In order to introduce flexible linkers and domains, we recall that we defined the protein in the model by its DNA binding interface. In this operational definition, we defined a protein unit of the binding interface as the part of the protein that contacts exactly 1 bp. Within this definition, it is not important to which protein domain this unit belongs.

For the characteristics of the complete protein–DNA contact, there is therefore no difference between, say, a protein that is made of two domains of 3 bp length but has only one domain bound and a protein that is made of only one domain of 3 bp which is bound.

A protein type with two domains can have either domain one, domain one *and* two, or only domain two bound. In the model, these different bound states have to be introduced as *different* proteins because their DNA binding interface differs. This is justified physically as well: Following the ergodic hypothesis, because the distinguishable states of the protein bound in either of the presented ways define *different* states of the system, we need to treat those states as being separate when we look at the system ensemble. Since we assumed that the chemical potential is independent of the state of the system as a whole, we can treat them as different proteins.

In order to avoid confusion, we choose to distinguish a specific protein for which we have already defined the binding interface domains by the appellation of the *version* of the protein; for example, a three-domain protein with a single domain bound is a *version* of the three-domain protein. We often do not care which particular domain binds or what is the order of the bound domains in a multidomain protein. Therefore, sticking to the example of three-domain protein, there is a set of versions with a single domain bound, and we call an instance of this set a *one-domain version* of the protein. In the implementation, we must keep track of the versions because the different possibilities define different system states, entering separately into the partition function. The ensemble average in the model must be taken over all *versions* of the protein not its *domain versions* (sets).

This nomenclature might sound complex but it conveys the correct physical point of view. Additionally, thinking of the proteins in this way facilitates the implementation enormously because the entities over which the sums in the partition functions run are exactly the ones described. A more precise and mathematical formulation of these concepts can be found in Appendix.

In order to give a more vivid representation of those terms, Fig. 3.6 contains graphical representations of the concepts.

4.3. Rules for multidomain binding

There are two major rules to discuss for the binding of several domains: In the model, they are required not to produce a gap on the DNA and furthermore not to permute the domain order.

The first condition is somewhat arbitrary because only little is known about the nonspecific binding properties of zinc finger proteins. In specific binding, though, it seems that neither for CTCF nor for ZiF268 large gaps occur, which would mean that the binding length of the protein on the DNA would enlarge considerably as well (see Ohlsson et al., 2001 for CTCF and Pabo et al., 2001 for ZiF268).

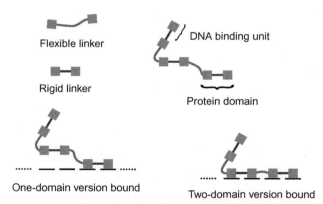

Figure 3.6 The terms *domain* and *protein domain version* as used in the model and their bound states are represented graphically. Note that the domain versions could have also different domains bound (e.g., the first two for the two-domain version). This would fall under the same name. (For color version of this figure, the reader is referred to the online version of this chapter.)

The condition that the order shall be preserved directly reflects the finite length of the linker which would have to span the distance over the zinc finger binding out of the order.

Every combination of domains obeying these rules is allowed. Some examples that illustrate which states are allowed and which are forbidden are represented graphically in Fig. 3.7.

4.4. States of the system with flexible linkers and domains

Without flexible linkers, it suffices to say by which unit of which protein type a DNA unit is bound to define its state. Now, we need to add the information by which version of the protein-type it is bound. If we have this information for all DNA units, the state of the DNA unit is defined. By looking at all *possible* states of the DNA and using the statistical physics formalism of the partition function (which changes strongly upon introducing flexible linkers), equilibrium values for physical quantities like the binding affinity of a protein for a given DNA unit can be obtained. A more mathematical description of this process can be found in Appendix.

The probability to find a certain protein-type bound with its first position to a DNA position is the sum of all probabilities for all versions of the protein being bound there with their first position. The binding affinity is

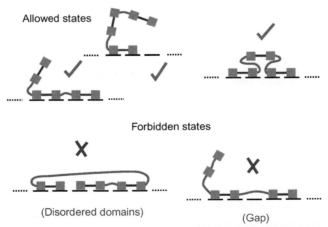

Figure 3.7 The figure shows examples of allowed and disallowed states in the model. Gaps on the DNA and disordered domains are forbidden, but binding of domains with dissociated other domains is allowed. Note that in the later implementations, one domain consists of a 4-bp binding interface. In this figure only a 2-bp binding interface is shown for clarity of presentation. (For color version of this figure, the reader is referred to the online version of this chapter.)

(by definition) calculated by taking the probability to find a DNA position bound by the first protein position and then adding it to all DNA positions that would also be bound in this case. This must be done individually for all protein versions because they differ in length.

The number of versions of a protein is called *cardinality* of the set of protein versions. Recall that every protein version corresponds to a conformation in which the protein can be bound to the DNA. The cardinality is important because it increases with the introduction of flexible linkers. This leads to an increase in binding affinity.

4.5. The effect of flexible linkers on the binding affinity

We expect three parameters to influence binding affinity in a system containing one protein type: The binding energy ε, the chemical potential μ, and the binding entropy of a state. With binding entropy of a state, we mean here a measure for the number of different states in which a protein can be bound to the DNA. For the model version of the three-zinc finger domain protein ZiF268, there are, for example, three protein versions that have one domain as the DNA binding interface. There are three protein versions that have two domains (one with the first and second domain, one with the first and third domain, one with the second and third domain); and there is one protein version that has all three domains as its DNA binding interface.

When there is more than one protein type, another entropic effect prefers shorter protein types. This can be intuitively understood with the observation that the more we restrict the system in phase space with a condition, the more entropy will work against this condition. The condition "long protein bound" restricts phase space more than "short protein bound." Less states of the whole system are compatible with the first condition, and it becomes less likely that it is true. For example, imagine first a chain with N units and a protein-type that has also a length of N units. Since we forbid overlapping binding, there is *one* state in which the protein is bound, spanning the whole DNA piece, and one in which it is not, the case of an empty chain. Next, imagine a chain with N units and a protein of length 1. There are 2^{N-1} states in which at least one protein is bound and one state in which nothing is bound.

This is complicated by flexible linkers because a long protein comprises many long *and* short versions (e.g., N versions with one domain when the protein type has length N). These short versions contribute strongly to the binding affinity.

This trend is limited by the chemical potential: Due to low concentrations, every additionally bound protein will produce a large negative free energy factor. The system cannot simply fill the lattice with little proteins until some combinatorial maximum of the states is reached. The free energy cost of introducing many proteins to the DNA would be too high. However, a nonzero binding energy gives weight to longer protein versions because they give a larger (negative) contribution to the state's overall binding energy while having by assumption the same chemical potential as small proteins.

One can envision that if we then add a position-dependent potential, things get complicated very quickly. This was the motivation to tackle the problem numerically.

The exact goal of the simulation was set to analyze the behavior of a zinc finger protein with three domains connected by flexible linkers in the presence of a sequence-dependent potential because for such a protein (ZiF268) we had data to compare the results of the model with. Two points were of special interest: What is the effect of flexible linkers on binding affinity? Do different domain versions (defined by different numbers of domains bound to the DNA) have a nontrivial distribution if we impose realistic parameters?

4.6. Enhancement of binding affinity of zinc finger proteins by flexible linkers

To distinguish the effects just described, we used two systems: A competitive and a noncompetitive one. The first one should show effects that stem from competition of different protein types, whereas the second one should not.

We did this by introducing into the first system only one protein-type and repeating the simulation for different protein types. The second system contained all protein types at once.

The protein types differed in the number of flexible linkers and therefore in the number of domains. The maximum length of all protein types was the same, 16 bp.

In the case of large chemical potentials (large in relation to the binding energy, which does not mean unrealistically large), the two system should converge because the competitive effects vanish.

The non-competitive system was also used as validation and a proof of principle as described in Appendix.

Without values for the parameters (the binding energy ε and the chemical potential μ) that have at least roughly the correct order of magnitude, the simulation would be futile: The behavior of the system might be strongly biased by consequences of unrealistic values, for example, when the binding affinity becomes too high.

Therefore, we searched for realistic parameters to introduce. The first problem was that it is hard to find a "nonspecific binding energy" in the literature, most probably because there is no such thing: Nonspecific binding is best described by a rough landscape with stochastic structure (Slutsky & Mirny, 2004). Indeed, the simulation could have been run several times with a stochastic position–dependent potential varying around some mean value, turning it into a Monte Carlo simulation, but we left that possibility aside for future investigations.

Thus, reasonable assumptions were made for the possible *range* of the binding energy, and then a parameter study was set up for this range. No estimation for the exact values for the binding energy or the chemical potential was made, just for the order of magnitude. This was justified from a broader point of view: As the claim is that binding affinity is increased by the introduction of flexible linkers, the claim would be weak if it would depend on the exact value of the parameters; we prove the stability of the effect by showing that the latter is visible for several orders of magnitude around reasonable values.

To start, it is clear that a nonspecific binding energy for a zinc finger protein should definitely be smaller than its specific binding energy which (Wolfe, Nekludova, & Pabo, 2000) gave in form of a dissociation constant between $k_d = 10^{-8}$ and $k_d = 10^{-10}$ M for Zif268. This corresponds to a value between -18.0 and -23.4 $k_b T$ for the whole protein and therefore between -1.5 and -2.0 $k_b T$ for each DNA-contacting unit of the zinc fingers, depending whether the zinc finger domain contacts 3 or 4 bp, a matter of dispute. The nonspecific binding energy must therefore lie substantially below this value.

Also, the measured dissociation constants should already *include* the effect we want to measure: If nonspecific binding facilitates specific binding, the measured values for nonspecific binding energy are *overestimated* in absolute magnitude.

In accordance with this order of magnitude, Slutsky and Mirny (2004) show that if the nonspecific binding energy were ≤ -2 $k_b T$ for the whole protein, no effective scanning of the DNA would be possible. Although we did not simulate sliding or in fact any dynamics directly, the argument still holds: If the energy for nonspecific binding was larger, the protein would spend prohibitively much time at the wrong positions in any search process.

For Zif268 that would mean that the binding energy per base contact must be larger (less negative) than -0.167 to $-0.22\,k_bT$. Altogether, we can take $-0.15\,k_bT$ as a rough lower bound for ε in the model.

Typical concentrations of transcription factors are in the range of nanomolar, which would indicate a chemical potential in the range of $20\,k_bT$. The simulation checked the range of $10\text{--}25\,k_bT$.

Note that, as we simulated a 16-bp protein in contrast to the calculations of ε (where we assumed ZiF268, a protein with 9 or 12 bp binding interface), we should have corrected the upper bound for its binding energy. Yet, the change in magnitude would have been only from -0.15 to $-0.12\,k_bT$. This is negligible regarding the fact that we set up a study for a whole range of binding energies.

Figure 3.8 shows the results for the competitive system and Fig. 3.9 the results for the noncompetitive system. Each system was simulated for the

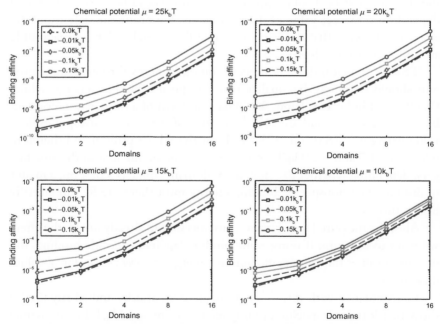

Figure 3.8 The competitive result. Binding of several 16 bp spanning protein with different numbers of domains as indicated. All proteins were in the same system. The plot shows their binding affinity (absolute probability to find a DNA unit bound by a protein) averaged over 20 bp in the inner region of the DNA chain with length $N=256$ bp. The binding energy per bound unit ε and the chemical potential μ were used as parameters as shown. (See Color Insert.)

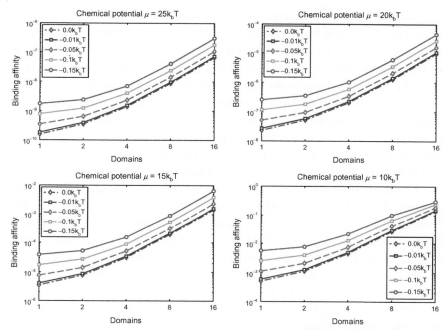

Figure 3.9 The noncompetitive result. Separate systems were set up, each of which contained one protein type with a 16-bp binding interface and number of domains as indicated. The plot shows the binding affinity (absolute probability to find a DNA unit bound by a protein) averaged over 20 bp in the inner region of the DNA chain with length $N = 256$ bp. The binding energy per bound unit ε and the chemical potential μ were used as parameters as shown. (See Color Insert.)

chemical potentials μ equal to 10.0, 15.0, 20.0, and 25.0 $k_b T$, while the binding energy per bound unit ε was set to -0.0, -0.01, -0.05, -0.1, and -0.15 $k_b T$, respectively. We stress again that all realistic values of ε should fall in that range: $\varepsilon > 0.0\,k_b T$ would be certainly overestimated because the interaction between the protein and the DNA (-backbone) is not repulsive; $\varepsilon < = -0.15\,k_b T$ was ruled out earlier.

The plots show the binding affinity, averaged over the middle 20 bp of a chain of length $N = 256$ bp for the different protein types of the same length but different numbers of flexible linkers and domains (indicated by the x-axis).

The central expectation that there is an increase of binding affinity with an increasing number of domains and flexible linkers is met.

This increase is stronger than linear partly because the increase of the cardinality of the set of protein versions with the number of domains follows an exponential law:

We can denote every version of a protein (as defined earlier) with N domains as a N digit binary number ($101100\cdots111$), where a 1 at the nth position states that the nth domain takes part at the DNA binding interface of this protein version, and 0 that it does not. The cardinality of this set is then the number of combinations of 1 and 0 s, which is simply the number 2^{N-1} (we excluded the "empty" protein). Therefore, the weighting factor corresponding, for example, to the case "protein type ξ bound with its first position at i" will be a sum with 2^{N-1} parts if the protein has N domains and because we sum over all protein versions. The binding affinity includes this factor because the probability is proportional to the weighting factor.

The increase in binding affinity is not exactly 2^{N-1} because the protein versions have different lengths and give different contributions to the weighting factor. Also the displacement of longer proteins by shorter ones influences the system even if one has only one protein (the noncompetitive system) because the different protein *versions* have different lengths.

As a last point, the definition of the binding affinity gives more weight to longer protein versions: The 16-bp protein version of a protein occupies 16 positions when bound and adds this probability to all positions when calculating the binding affinity.

4.7. Effect of flexible linkers on the structure of binding maps of zinc finger proteins

Next we looked at a protein with three zinc finger domains binding to a chain in the presence of a sequence-dependent potential of varying height. The main quantities of interest were the binding affinity maps (the absolute probability to find a protein bound for a specific DNA position calculated for the whole chain) for the different domain versions of the protein (defined by the number of domains that bind the DNA).

The potential was set in the middle of a long chain and had a simple rectangular form. A discussion of the form and the height can be found in Appendix.

The results are shown in Fig. 3.10, where different colors indicate the different domain numbers. These comprise different protein versions: In "two Domains," for example, three protein versions are included: the version with domain one and two bound, the version with domain one and three bound, and the version with domain two and three bound.

The ratios between the bulk binding energy per bound unit (which a DNA–protein contact contributes to the free energy outside of the

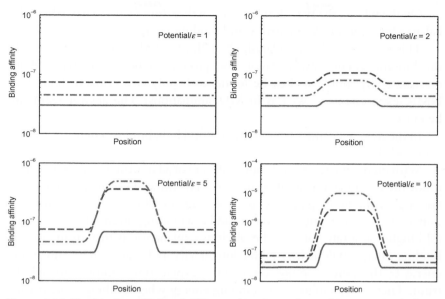

Figure 3.10 The binding affinity of different domain versions of a protein with three domains in the presence of a square potential. The potential had length $L = 40$ bp and was centered around position 128. The domain versions are defined by the number of domains with which they bind the DNA, as shown. Potential/ε denotes the ratio of the nonspecific binding energy ε and the height of the square potential. The data are shown for a chain of length $N = 256$ bp and $\varepsilon = -0.05$ k_bT. (For color version of this figure, the reader is referred to the online version of this chapter.)

potential) and the potential height, which serve as a parameter, are indicated above every plot.

Competitive effects can be neglected as the absolute binding probability of a DNA unit lies with the given parameters at 10^{-8}–10^{-7}.

For no potential (ratio $= 1.0$), all domain versions have even distributions of different height.

The two-domain version has the highest binding affinity. This domain version has a higher binding energy than the one-domain version and more ways to be bound to the DNA than the three-domain version. The three-domain version has the highest binding energy but only one way to bind to the DNA. It has a lower binding affinity than the two-domain version but a higher one than the one-domain version. The one-domain version has the lowest binding affinity, although it has the same number of possibilities to bind to the DNA as the two-domain version.

On the whole, the two-domain version therefore seems to have the right balance between binding entropy and binding energy.

This is exactly what Zandarashvili et al. (2012) describe as a surprise: If the binding energy always favors the protein to be bound with all its domains, why should it then be found with only two domains bound in NMR experiments as well as in MD simulations? The answer seems to be entropy.

In the next plots, the ratio of potential and bulk binding energy is given a value >1.0.

With increasing ratio, the binding affinity of all domain versions increases inside the potential, but at different paces. The three-domain version has the fastest growing binding affinity in the potential region. It surpasses the two domain version at a ratio of about 5 and has a binding affinity three- to fourfold higher at a ratio of 10. The obvious explanation is that an increasing binding energy will always prefer longer factors because they gain more energy by the increase. Note, yet, that this explanation cannot be the whole truth: There is no difference in binding energy whether three protein versions bind with one domain or one protein version is bound with three domains. The former case is yet disfavored by the chemical potential, which therefore must also be a major factor in the distribution.

It is very interesting to note that these effects begin to play a role exactly in the parameter region that is assumed to be at the limit of nonspecific binding. It would be tempting to interpret the three-domain version as the protein bound to its specific site, but as we elaborated earlier, specific binding with its topological implications is far out of reach for the model. Though we can securely state that the three-domain version nonspecifically bound to the DNA is topologically more similar to the specifically bound complex as a domain version with less domains.

5. A METHOD TO COMPUTE SIMILARITY FOR UNSTRUCTURED PROTEINS

As it was discussed and proved in the previous chapters, CTCF is a flexible protein. Then, it is logical to ask how we treat these flexible proteins. How do we compare them to other structures? How does it bind to different structures? Can we predict potential binding sites? It is obvious now that the latter is not a lock and key problem anymore. The following section aims to introduce a very engaging method which promises to answer all these questions.

First, we summarize the present generic geometric treatment of proteins which presumes that proteins are crystalline structures. Then we introduce a new approach to the problem and illustrate its potentials with a few examples.

5.1. Geometric similarity

Let us now shortly describe how geometric similarity can be determined. Our aim here is not to exhaust the meaning of geometric similarity, instead we would like to provide a simple basis for the comparison with our approach.

In order to determine the similarity of chemical structures, first we need to find an adequate shape representation of the molecules. Calculating geometric similarity practically means calculating the similarity of shapes of the molecules. The shape of chemical structures can be estimated by placing a sphere to the center of the structure's each atom with a radius corresponding to the atom's Van der Waals radius. These spheres can be Gaussian surfaces but in the most unsophisticated case, we would use simple hard spheres.

A very important step when estimating geometric similarity is the calculation of the best alignment of the structures as geometric similarity measures heavily depend on the orientation of the shapes. In other words, this means that no matter what the definition of the geometric similarity is, we need to maximize it over all rotations and translations of one of the structures while keeping the other one fixed. This is computationally a very expensive process.

After we achieve the best alignment, we can proceed with the last step of calculating geometric similarity. For this reason, we overlay the shapes and calculate the section and union of volumes as it is illustrated in Fig. 3.11 with the help of adenine (A) and guanine (B).

The Hodgkin measure of geometric similarity (Willett, Barnard, & Downs, 1998) is defined as the Tanimoto measure (Rogers & Tanimoto, 1960) of the volumes, that is, the volume of the section over the volume of the union. In a more general case, the volume can be replaced with other descriptors; however, we constrain ourselves to calculating only volumes as it already provides the necessary base for comparison.

A very good and detailed review of geometric similarity measures can be found in Willett et al. (1998).

5.2. Geometry versus topology

As it was pointed out, geometric similarity measures focus on how the actual shape of the molecules compare. However, as the flexible linker model also captures this correctly, a vast amount of the proteins we encounter in living

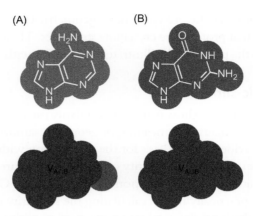

Figure 3.11 Geometric similarity is calculated as the ratio of the volumes of the section ($V_{A \cap B}$) and the union ($V_{A \cup B}$) of the shapes representing the chemical structures. (For color version of this figure, the reader is referred to the online version of this chapter.)

organisms are very flexible structures. This flexibility allows them to frame to different constraints or to other shapes. On the other hand, the capability of such adaptability automatically triggers the fact that there must be an almost unlimited number of shapes, an ensemble of shapes, a protein could achieve. Some of these shapes are of course similar to each other, but some of them might be totally different. If we randomly pick two shapes from the ensemble, odds are high that the geometric similarity measure indicates a reduced similarity between the two. This is of course correct if we are talking strictly about geometry. However, proteins are flexible for a good reason: to be able to adapt and bind to different shapes. Thus, when talking about similarity measures as means of identifying potential bindings, we must not neglect the capability of a protein of changing its conformation. We should be able to tell that the two conformations came from the same ensemble, that is, they represent the same protein or, at least, we should be able to indicate that small scale (local) structures are built up in the same way. This means that structures like bonds, rings, loops, etc. must be preserved. In a more mathematical language, we would say that the *topology* of the two systems must be the same.

We have introduced here the key concept of our method: topology. To briefly summarize, topology is the field of mathematics which studies properties of objects which are preserved under certain deformations like stretching and bending. Note that these deformations are exactly the ones that allow a protein to change its shape. Thus, if we look at the topology of proteins instead of their geometry, we should be able to decide whether two instances are chosen from the same ensemble or not.

5.3. Using topology to compute similarity

Looking at the topology of a given molecule carries a huge advantage. The topology of a structure is invariant with respect to similarity transformations. This means that for a given object no matter what the point of reference is, the topology is the same. Thus, by looking at the topology of the objects, one avoids the expensive calculation of the best alignment.

When it comes to investigating the topology of a protein, we can think of a few different approaches. One is to treat the proteins as graphs—abstract mathematical objects consisting of nodes interconnected by edges or links—in which the nodes would represent the atoms and edges the chemical bonds. Afterward, we could forget about the protein itself and could carry on looking only at the graph. We can pick from a vast amount of well-established and studied measures introduced by experts in the field which would characterize the topology of the graph. However, in this case, we have to be careful as these measures are usually independent of the physical size of the investigated system, that is, these measures are invariant with respect to the rescaling of the system. In the present situation, this behavior is not desired.

Another approach is to take advantage of the recent developments of computational geometry as it promises suitable techniques crafted specifically to analyze topological properties of a given object by calculating so-called *topological invariants*. The topological invariants we will focus on are the number of connected components, the number of holes, and the number of voids in the investigated object. Connected component in this context means parts which are connected to each other. For example, a regular ball has a single connected component (since it is a single piece), no holes and a single void (inside). A piece of paper also has a single connected component, no holes and no voids. If we tear the paper in two, the system composed from the two (now separated) pieces of paper have two connected components (since the two pieces are not connected to each other anymore), no holes and no voids. If we take a pencil and poke a hole in one of the papers, we end up with a system with two connected components, one hole and no void. In the field of topology, the mentioned topological invariants correspond to the so-called *Betti numbers* (Carlsson, 2009). The number of connected components is the 0th Betti number, the number of holes is the 1st Betti number while the number of voids is the 2nd Betti number.

If we try to compare two objects relying only on the number of connected components, holes and voids, we would quickly run into trouble. For instance, we would find that a regular coffee cup (one component, one hole in the handler and no void) is similar to a donut (one component, one hole in the middle and no voids).

5.3.1 A barcode representation of topology

In order to avoid such blunders when comparing proteins, we resort to the following abstraction. We remove all the bonds from the proteins keeping only the atoms. From now on, we are not interested in the physical meaning of atoms, thus we will refer to them simply as points. Then we adopt the following procedure:

- we define a distance scale d
- we connect all point pairs that are closer than d with a line
- we calculate the topological invariants for the obtained structure

We repeat these steps for different d values and follow how the topological invariants behave as we vary d. To do this consistently, we define a minimal and a maximal value for d and fix the number of values d will have between its extremities, that is, we divide the range defined by the minimum and the maximum to equal intervals. First, we set d to its minimal value, register the values of the topological invariants at this value then increase d to its next allowed value. We repeat these steps until we exhaust the allowed values for d.

At this point, it is a valid question what is counted as a hole and how voids can form when all we do is connecting points with lines. The answer lies in the definition of topological building blocks which are points, lines, triangles, tetrahedrons, and their higher dimensional analogs. According to this, triangles do not count as holes, instead they constitute faces. Similarly, the space enclosed by a tetrahedron is not counted as void. Thus, any polygon which is not a triangle constitutes a hole; similarly, any polyhedron which is not a tetrahedron contains a void. Note that the faces of the polyhedrons can only be triangles—otherwise there would be a hole in the wall of the polyhedron, thus it would not be a polyhedron anymore.

After we scan the system with the procedure described earlier, we know the numbers of connected components, holes, and voids for each value of d. The acquired information can be summarized in a diagram in the following way:

- each instance of connected component, hole, and void will be represented by a bar
- the starting point of the bar will correspond to the value of d when the instance came into existence

- the end point of the bar will correspond to the value of d at which the instance ceased to exist

The bars for connected components are somewhat special as connected components unite as d increases. This process can be viewed as one of the connected components embeds the other. Accordingly, the bar of the embedded component will end at the point where the component was embedded, while the bar of the embedder component will continue until the letter is embedded in another component. The role of embedded and embedder is arbitrary. It is easy to see that one of the bars for connected components will persist even at the highest values of d as there will always be at least one connected component, thus this bar can be neglected as it does not carry any information.

The diagram compiled in the previously described way will be a barcode representation of the topology of the system in which each bar represents the interval of d over which the corresponding topological feature persists (persistence interval). An example for such a barcode can be seen in Fig. 3.12. This representation was developed by Carlsson and his collaborators and a

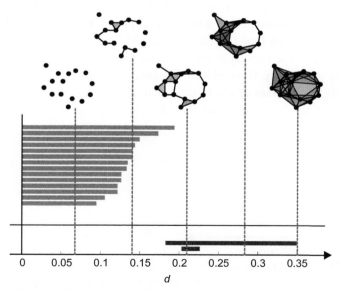

Figure 3.12 Barcodes for a given set of points in 2D. The horizontal axis represents the distance parameter. Persisting "features" are arranged on the vertical axis in an arbitrary order. On the top of the figure, the procedure of connecting points is illustrated for a few values of d. Shaded faces in these illustrations signal formed triangles. Each triangle has a different color. (For interpretation of the references to colour in this figure legend, the reader is referred to the online version of this chapter.)

very good review of their work can be found in Ghrist (2008). Betti numbers and calculating persistence intervals are discussed in more detail in Appendix.

5.3.2 A barcode-based topological similarity measure

So far, we described how to represent the topology of proteins with barcodes. Since these barcodes encode the topology, by comparing them, we actually compare topologies. A very important extra feature of the barcodes is that while they encode topological invariants, they do not neglect the physical configuration of the proteins as they carry information about distances. In principle, it would be possible to reconstruct a configuration from its barcode representation; however, this would require solving an optimization problem. Thus, comparing objects by looking at their barcode representation should correlate with measuring their geometric similarity. In addition, it should be possible to match geometrically different but topologically similar structures.

Comparing barcodes to each other requires a definition of a distance measure. For this purpose, we need to remember that each bar can be characterized by two numbers: its start and end point. This means that we can represent the barcodes for a given topological invariant as points in a 2D plot in which one of the axes represents the start points while the other axis represents the end points. If we plot for instance two sets of barcodes for holes, we would get a diagram similar to the one in Fig. 3.13A. Then for each point from one of the sets, we could find the closest point from the other set (Fig. 3.13B).

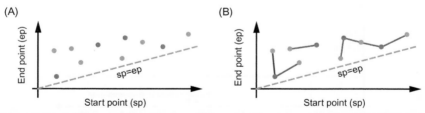

Figure 3.13 A different representation of barcodes. The start and end values of the bars are represented on different axes. Each bar corresponds to a point. Note that all points should be above the line which is defined by the start values being equal to the end values since each bar ends later than it starts. Blue and magenta points correspond to different shapes. The red and the green lines connect closest points of the different sets. The red line indicates the largest distance between the closest points. (For interpretation of the references to color in this figure legend, the reader is referred to the online version of this chapter.)

We measure the distances between the pairs of closest points. Choosing the largest distance as a similarity distance assumes a worst-case similarity distance of the sets (in set theory, this would be called the Haussdorff metric; Rockafellar & Wets, 1998). Then, our topological similarity distance (TSD) of the topologies will be defined as the sum of these largest distances over the different topological invariants (connected components, holes and voids). Note that what we just calculated is a sum of distances which is not a proper distance measure; we call it a distance to emphasize that smaller values will indicate more resemblance. In order to have a similarity measure, we will use TSM = 1 − TSD. Then, values of TSM close to one mean very similar topologies, while smaller values indicate less similar topologies. A more precise definition of this similarity measure is found in Appendix.

5.4. A comparison between geometric and topological similarity

In order to verify the previously defined topological similarity measure, we tested it on a benchmark database and compare it against the Hodgkin geometric similarity measure. For this purpose, we chose the DUD (Huang, Shoichet, & Irwin, 2006) database. We present the results for two target proteins: Acetylcholinesterase (AChe) and Thymidinkinase (TK). The DUD provides 105 ligands and 3732 decoys for AChe while for TK we have 22 ligands and 785 decoys. We divide each ligand and decoy in 2–3 subsystems depending on their size and homogeneously sample neighborhoods of similar sizes from AChe and TK. We compare each ligand and decoy to its corresponding target by comparing all possible combinations of pairs of neighborhoods between them. Topological invariants are calculated with the freely available *perseus* software (Nanda, 2012), while geometric similarities are calculated with an in-house developed library. We mark the results of the comparison on a plot with the x-axis representing the topological similarity measure, while the y-axis represents the geometric similarity. Since on these plots many of the points overlap, we calculate the density of the points and present them as a density plot.

It is already clear by inspecting Fig. 3.14—the plot for the AChe target—that the topological and geometric similarity measures do correlate, as expected. We can also see that while the geometric similarity ranks relatively low for some of the neighborhood pairs, they might still have very similar topologies. This suggests that the introduced topological similarity measure is able to pick up additional potential binding sites, which is a very important achievement over the geometric similarity measure (lower right side of the

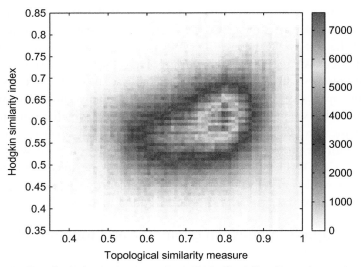

Figure 3.14 Topological versus geometric similarity for AChe. (For color version of this figure, the reader is referred to the online version of this chapter.)

figure). Similar observations can be made from Fig. 3.15, which presents the results for the TK target.

In order to prove that our similarity measure is consistent and well behaving, we calculate the distribution of the highest similarity measure values for the decoys and the ligands separately and expect a similar distribution for the two.

Examining Figs. 3.16 and 3.17, we can deduce that our expectations are satisfied. Thus, we can conclude that the presented method works well on real-life scenarios and correlates with the geometric similarity, which was an expected behavior. Moreover, our method has a huge advantage over the geometric similarity measure: it can indicate additional potential binding sites which were totally neglected by the geometric similarity measure.

5.5. An application to CTCF

As a first test-application on CTCF we considered several configurations of the protein achieved by bending some of the linkers and then relaxing the system. In our test, we chose the subunits composed by the bent linkers and the two zinc finger domains at the end of the former and attempt to demonstrate that although these units changed their geometry, we are still able to show that they are similar. The units are illustrated in Fig. 3.18.

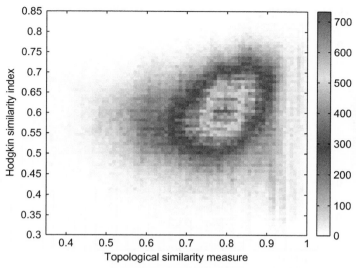

Figure 3.15 Results for the topological and geometric similarity measures for TK. (For color version of this figure, the reader is referred to the online version of this chapter.)

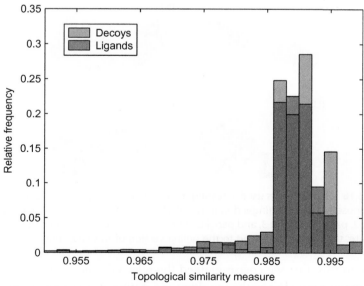

Figure 3.16 Distribution of the highest values of the similarity measure in case of AChe. (For color version of this figure, the reader is referred to the online version of this chapter.)

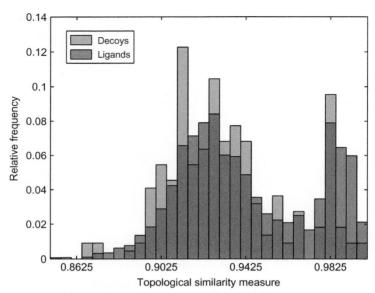

Figure 3.17 Distribution of the highest values of the similarity measure in case of TK. (For color version of this figure, the reader is referred to the online version of this chapter.)

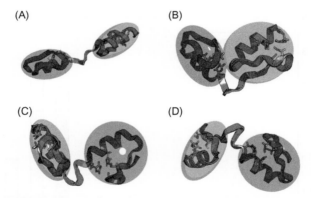

Figure 3.18 CTCF subunits used in testing geometric versus topological similarity. They are composed of two zinc finger domains and a flexible linker. Panel A subunit 1, panel B subunit 2, panel C subunit 3, and panel D subunit 4 (see Table 3.1). (For color version of this figure, the reader is referred to the online version of this chapter.)

To do this, first we coarse-grained the selected units, as proper coarse-graining must not change the topology and it improves computation. Then we calculated the topological and geometric similarity measures. Results are summarized in Table 3.1.

Table 3.1 Results for the geometric and topological similarity comparison of CTCF subunits

Tests	Hodgkin similarity	Topological similarity
Subunit 1 versus subunit 2	0.532	0.98847
Subunit 1 versus subunit 3	0.641	0.99012
Subunit 1 versus subunit 4	0.602	0.99012
Subunit 2 versus subunit 3	0.669	0.98378
Subunit 2 versus subunit 4	0.667	0.98413
Subunit 3 versus subunit 4	0.955	0.99997

Although geometric similarity indicates differences, it can be seen that topological similarity clearly shows that the domains have similar topologies.

6. DISCUSSION AND OUTLOOK

In this chapter, we have focused on the flexibility of the linkers as a key aspect of CTCF. Because of the flexibility, CTCF can be considered a flexible polymer with a monomer size slightly larger than a zinc finger. Thus, as a consequence, entropy will be very important for the binding. From this finding, we have derived a simple model for the binding.

It has been shown that even in a very simple model for the binding of zinc finger domains connected by flexible linkers, a structural change of the nonspecific binding of a ZiF268-like protein appears in regions of increased binding energy. The binding mode is determined by the balance between entropy, that is, by the number of states the protein can use to bind to the DNA and the binding energy this state has. In regions of low binding energy, the binding affinity is maximal for the protein binding with two domains. Inside an area of increased binding energy, the state with three domains bound becomes the most likely when the increase in binding energy is about fivefold. Increasing this value further results in a stronger effect and at a ratio of the binding energy inside and outside of about 10, the probability to bind with three domains is about fourfold higher than to bind with two domains. Figure 3.19 represents this general distribution of the states inside and outside qualitatively.

This is remarkable because it points to a strong microscopic effect driven by entropy that becomes apparent even in the basic model.

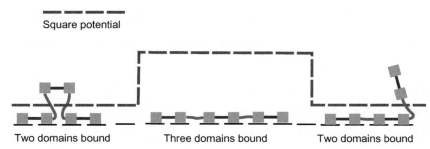

Figure 3.19 A representation of the square potential used and the states that have been observed to occur inside and outside of this square potential. Note that in the actual simulation, zinc finger domain was assumed to contact 4 bp on the DNA, while in this graph only a 2-bp interface is depicted for clarity of representation. (For color version of this figure, the reader is referred to the online version of this chapter.)

Although the mechanisms of facilitated target search are under investigation since more than four decades and are now a standard topic in biophysical textbooks, the theory and its various extensions are far from being generally accepted. In fact, light has been shed upon a new array of critical aspects of theory and experiment in recent years (Kolomeisky, 2011).

Therefore, it is sensible to formulate the interpretation of the results first, independent of the terminology of this theory. Interpretations pointing in this direction may then be given.

The main question is what difference it makes for a ZiF268-like protein whether it is bound with more or less domains (two or three for ZiF268). First of all, it is obvious that a state which is bound with three domains is less flexible on the DNA because it has more topological restrictions. Furthermore, on average the extent of this state is longer than the extent of a state bound with two domains. Both assumptions may result in a higher binding affinity for the latter state (that is, the state with two domains bound). Also, the state with three domains bound resembles topologically the specifically bound state which is also bound with three domains for ZiF268 (Zandarashvili et al., 2012). Yet, the extent of this resemblance is not known. It has been shown that the binding of zinc finger proteins to their specific site is connected to changes in their intrinsic structure, which have no counterpart in the nonspecific binding (e.g., structuring of the linker region; Hyre & Klevit, 1998).

The states bound with two domains in regions of low binding energy are bound more transiently. They should have a greater propensity to dissociate and associate again with a different domain combination. Also, the probability for them to dissociate and diffuse away completely from the original binding point is higher, but in regions of low binding energy this is compensated by the higher binding entropy according to our results.

The freely floating zinc finger domains have been shown to have additional functions: In a system densely packed with DNA, the free zinc finger domain can engage in an interaction with another DNA chain and initiate a move of the whole protein from one DNA chain to the other, which has been called "monkey bar mechanism" (Vuzman, Polonsky, & Levy, 2010).

The results as described are confirmed by experiments (Zandarashvili et al., 2012). In NMR and MD simulations, it was shown that when engaged in target searching, the ZiF268 protein is rather bound with two (seldom one) domains in contrast to the specifically bound complex where all three domains are in contact with the DNA. The free domain can then take part in the monkey bar mechanism.

We cannot interpret the state with three domains bound as the specifically bound state because zinc finger proteins exhibit topological changes when binding specifically. Yet, we could interpret this state as a precursor state to the specifically bound state. The probability to switch from this precursor state to the specifically bound state can be assumed to be high because all three zinc fingers are already on the DNA (as in the specifically bound state). We want to stress here that this result is principally independent of the concept of 1D versus 3D diffusion. The model does *not* simulate dynamics. The distribution samples static states. The way they are transformed into each other does not matter. The distribution would not change if, for example, we assumed a system exhibiting much 1D diffusion or if we assumed the total absence of it.

One point still needs mentioning: Zandarashvili et al. (2012) show that in the two-domain state, the probability to take part in the DNA binding is much higher for two special zinc fingers of ZiF268 than for the third. This is what he calls the asymmetrical roles of the zinc finger domains of ZiF268. This is a clear sign for some intrinsic asymmetry of the protein which is not captured by our model. There, the domains are absolutely equal. It would be possible to implement such an asymmetry by a change of the effective binding energy for only one domain, but that would be hard to justify. Where this asymmetry in ZiF268 comes from is not known, nor whether other zinc finger proteins show a similar behavior. To sum up, we can say that the experiments are in agreement with our model, but it seems that effects resulting from the topology or binding behavior of special proteins might have a substantial influence as well.

The theory of Facilitated Target Search has experienced severe attacks recently (Kolomeisky, 2011). Therefore, we have chosen to detach the interpretation in this context from the main results (which are independent of this theory).

Winter, Berg, and Von Hippel (1981) already noted in 1989 that the 1D–3D diffusion model has a problem when one takes into account the energy landscape for nonspecific binding): If this energy is not high enough, the system will not exhibit enough 1D diffusion to stay within the limits of the theory. On the other hand, if it is high enough, the average probing time of a position on the DNA gets too long to be productive. This has been termed the "Speed-Stability Paradox" by Slutsky and Mirny (2004). He also proposed a model that can solve this paradox theoretically: Search and Recognition states. The protein is supposed to exhibit conformational changes in an area of higher binding energy, a transition from a loosely bound state that diffuses quickly and has a low propensity for recognizing the correct specific binding site to a more tightly bound state that diffuses slowly but has a high propensity for recognizing the correct binding site.

The solution links molecular properties of the protein to the overall process but has the problem that the microscopic details of these states and the switch mechanism are a mystery.

In the light of our results for ZiF268, though, there is an easy interpretation for this protein: In regions of higher binding energy, there *is* a conformational switch that occurs only there, the resulting conformation should exhibit slower 1D diffusion, and in this conformation, it *is* more likely to recognize the specific binding site: the switch of having three domains bound (the "Recognition State") in contrast to two (the "Search State"). The state with two domains bound should have a higher 1D diffusion coefficient and even other modes of accelerating the diffusion (like monkey bar mechanisms) are known. This fast diffusing state appears in the (vastly dominating) area of an average specific binding energy and is less likely to recognize the specific binding site because it is topologically less alike to the specifically bound state. We cannot say much quantitatively because the increase in the diffusion coefficient would need to be simulated in a totally different simulation context than our model, but the qualitative agreement of our results with the considerations of Slutsky is remarkable.

The main point in which this interpretation differs from the one given before is that it makes statements about the 1D diffusive behavior of the different domain versions, which is a central point to the Berg-Winter model of Facilitated Target Search. The preceding interpretation, on the other hand, did not assume such diffusion.

Yet another aspect of the flexibility of the protein is that the conventional methods to determine similarity, protein–protein binding, or binding of small molecules need a revision. Here, we have presented one possibility in this direction: Instead of looking at geometric similarity we have proposed and shown it is viable to use topology. Of course, this is just the first step. Much more work needs to be done on similarity measures for flexible proteins like CTCF.

ACKNOWLEDGMENTS

We would very much like to thank Roel van Driel and Mariliis Tark-Dame for stimulating discussions. Furthermore, G. M. would like to thank the Research Training Group (RTG) 1653 Spatio/Temporal Graphical Models and Applications in Image Analysis, the Heidelberg Graduate School of Mathematical and Computational Methods for the Sciences, and the Institute for Theoretical Physics for funding. D. W. H. would like to thank Yixue Li and Lei Liu as well as the SIBS for discussions and hospitality.

APPENDIX

A.1. A model for zinc finger protein–DNA binding with flexible linkers: Energy and calculation of averages

The energies used are in units of $k_b T$.

The energy of the system is obtained by adding all energies of all DNA–protein contacts. Thereby, we set the energy of an unbound DNA unit to zero. The energy of a DNA–protein contact can be position-dependent (reflecting the DNA sequence preferences of proteins) and a non-zero chemical potential must be taken into account for every bound protein (*not* for every bound unit).

The complete change in free energy upon binding to position M for a protein is thus represented by

$$\Delta G_i = \mu + \sum_{i=M-\ell+1}^{M} \varepsilon(i), \qquad (A.1)$$

where M is the DNA position that is bound by the *first* unit of the protein, ℓ denotes the length of the binding interface, and $\varepsilon(i)$ the position-dependent binding energy per unit for this protein.

If we assume for clarity of presentation that only one kind of protein (but in several copies) binds to the DNA piece we want to model, we can denote by $\{M_s\}$ the set of all DNA positions that are bound by the *first* position of the protein in the state s and add up all contributions of Eq. (A.1) to the weighting factor w_s of this state s,

$$w_s = \sum_{\{M_s\}} \exp\left(\mu + \sum_{i=M_s-\ell+1}^{M_s} \varepsilon(i)\right). \tag{A.2}$$

The partition function is then simply given by the sum of those weighting factors of Eq. (A.2) over the set of all legal states $\{s\}$:

$$\mathcal{Z} = \sum_{\{s\}} w_s = \sum_{\{s\}} \sum_{\{M_s\}} \exp\left(\mu + \sum_{i=M_s-\ell+1}^{M_s} \varepsilon(i)\right). \tag{A.3}$$

A valid state is one that complies with the basic rules stated in the main text. If there is more than one protein type present, we need to sum over all protein types (that can differ in length). If flexible linkers are introduced, we need additionally to sum over all protein *versions*, that is, all possible DNA binding interfaces of the protein types.

The probability to find a certain condition fulfilled in the system is the ratio of the sum of all weighting factors of states fulfilling the condition and the complete partition function in Eq. (A.3). If we write, for example, the sum of all weighting factors of all states in which the position P is bound by the first unit of a protein with $\mathcal{Z}^P$, the probability of finding the position P bound by the first position of a protein is simply $\mathcal{Z}^P/\mathcal{Z}$. The binding affinity for the protein, which is defined as the probability to find the position on the DNA bound by *any* position of a protein, is then

$$p_i = \sum_{P=i}^{i+\ell-1} \mathcal{Z}^P/\mathcal{Z}, \tag{A.4}$$

where ℓ is again the length of the protein.

A.2. Setup and results

A.2.1 Implementation

We present the basic idea to calculate the partition function for the systems. We used a recursion relation as developed by Delisi (1974) with dynamic programming. The central observation is that if we look at the last unit of the DNA polymer, we can write down *excluding* possibilities for it: It can be either bound or free. If bound, it can be bound by different protein types, but only by one at a time. If bound by a certain protein type, it can be bound by different versions, but only by one at a time. Also, because we forbid overlapping binding, it can be bound only by the end of a protein. The partition function is the sum of all these excluding possibilities and can be written as

$$\mathcal{Z}(N) = \mathcal{Z}(N-1) + \sum_{\{\xi\},\{\gamma\}} e^{w_\gamma^\xi(N)+\mu} \mathcal{Z}\left(N - m_\gamma^\xi\right), \qquad \text{(A.4)}$$

where $\{\xi\}$ is the set of all proteins in the system, $\{\gamma\}$ is the set of all versions of the protein ξ, and m_γ^ξ is the corresponding length in base pair units. For convenience, we abbreviated the binding energy of the complete protein version ξ of type γ at position N as $w(\xi/\gamma)(N)$.

The chemical potential lacks indices because its value was taken to be the same for all versions of all proteins. The only distinction for the proteins in the set ξ was the number of linkers, and it is doubtful that this can be generally correlated with the chemical potential.

To calculate the partition function itself, the conditions

$$\mathcal{Z}(N) = \begin{cases} 1 : N = 0 \\ 0 : N < 0 \end{cases} \qquad \text{(A.5)}$$

are used. The first condition defines that the weighting factor for an empty system is 1 and the second that no negative chain lengths are allowed. The numerical value of 1 was chosen for simplicity. It cancels out when physical quantities are calculated, as presented now.

In order to calculate the binding map for a protein type ξ, the conditions

$$\mathcal{Z}^i(N) = \begin{cases} 1 : N = 0 \\ 0 : N < 0 \\ 0 : i = \text{not bound by first position of } \xi \\ \sum_{\{\gamma\}} e^{w_\gamma^\xi(i)+\mu} \mathcal{Z}\left(i - m_\gamma^\xi\right) : N = i \end{cases} \qquad \text{(A.6)}$$

were used, where $\mathcal{Z}^i$ is the reduced partition function subject to the condition that the protein type under observation is bound with its first position to the DNA position i. The third condition defines that position i must not be bound by anything else than the first position of protein type ξ and the fourth continues the recursion with the correct weighting factor if it is bound by the first position of a version of ξ. The probability of finding such a state is then $\mathcal{Z}^i/\mathcal{Z}$. An analogous equation can be easily written down when one searches the probability to have a certain domain version bound, that is, a certain number of domains.

A.2.2 Validation and proof of principle

As a first proof of principle and as a validation of the implementation, several noncompetitive systems were set up, each containing only one protein type. This protein type had a DNA binding interface of 16 bp but different numbers of flexible linkers and therefore different numbers of domains in the different systems:

The linker numbers of 0, 1, 3, 7, and 15 of the proteins divided the 16–bp binding interface of the protein into 1, 2, 4, 8, 16 domains of length 16, 8, 4, 2, and 1 bp. They were all added to *separate* systems, so this part corresponded to the noncompetitive setup described in the text. The systems did not interact in any way.

In the search of a proof of principle, we did not wish to impose realistic parameters, which were done later. We set $\mu = \varepsilon = 0$ $k_b T$ and therefore the weighting factors in 7 to 1 for all states. A physical interpretation would be that we looked at the case $T \rightarrow \infty$.

Figure 3.20 shows the results for this calculation. For validation of the implementation, two data sets are plotted: One represents the direct

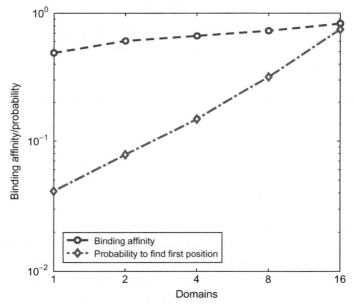

Figure 3.20 Noncompetitive proof of principle: The probability to find the first position bound and the binding affinity (the probability to find *some* position bound), averaged over the middle 20 bp of a chain of length $N = 256$ bp; the one protein type in all systems was 16 bp long but had a different number of flexible linkers and domains. Domain number is indicated by the x-axis. (For color version of this figure, the reader is referred to the online version of this chapter.)

outcome of the implementation, the probability of finding the first position of a protein bound to a certain DNA position. This is averaged over the inner 20 bp of the chain. The second curve shows what we have defined as the binding affinity: The cumulative probability of finding the DNA bound by *some* position of the protein, also averaged over the inner 20 bp. The different systems are indicated by the number of domains the proteins had in them (shown on the *x*-axis).

The binding affinity is, as expected, a generally higher number than the probability to find the first position of the protein. This is because for every position, the former is basically the latter summed over more than one position. For many flexible linkers the difference vanishes, what could have been expected: For a 16-bp protein with 15 linkers, that is, 16 domains, there are many short versions (e.g., 16 versions with 1 bp length, but only one version with 16 bp length); but the probability to find the first position of a 1-bp spanning version of a protein *is* the binding affinity for this protein version. Therefore, if the protein has many flexible linkers and therefore many short versions, the probability to find its first position bound gets similar to the binding affinity.

The central expectation that there is an increase in binding affinity with an increasing number of domains and flexible linkers is met.

A.3. Form and height of the potential

The position–dependent potential is hard to map to a real cellular system. It is clear that every function that can be implemented must be artificial because the binding energy depends on the exact nucleotide sequence (and even that is a coarse simplification leaving out many effects, e.g., 3D structure). Therefore, we state here explicitly that we did not want to simulate a specific potential but an area of increased binding energy.

The important characteristic of this potential is that the energy of a protein–DNA contact is more negative inside than outside it. How much more negative is a question difficult to answer. Therefore, we enacted a parameter study for several values for the ratio of binding energy inside and outside. With the chemical potential set to 20.0 $k_b T$ and the binding energy outside the potential set to -0.05 $k_b T$, the ratio of outside/inside binding energy was set successively to 1, 2, 5, and 10. These values were chosen such that the binding energy inside the potential covers roughly all values from -0.05 $k_b T$ to the lower bound for specific binding energy (as estimated in the main text).

The potential was chosen to have the form of a square potential. The definition of the inside of the potential as a piece of the DNA with an increased effective binding energy and the consideration that a form as simple as possible is preferable for a proof of principle suggested that it is convenient to use the simplest form accessible. Its width on the chain of length 256 bp was chosen to be 40 bp, and it was centered around position 128 to keep the region of interest free of border effects.

A.4. Numerical setup

In order to set up the simulation, the recurrent relation algorithm was implemented as a C++ program. A position-dependent potential was included, as well as the ability to calculate the binding maps for proteins and domain versions (defined by how many domains of a protein bind to the DNA). With dynamic programming implemented, the calculation of all data presented in this chapter took less than an hour on a standard PC (as of 2012, 2 GHz, 4 Gb RAM).

The zinc finger domain was generally assumed to contact 4 bp on the DNA chain.

A.5. Betti numbers and persistence intervals

As it was mentioned, the first three Betti numbers represent the number of connected components, the number of holes and the number of voids. Of course the concept of Betti numbers is more general and not limited to three-dimensional spaces. Betti numbers are defined for general topological spaces and are standard tools in algebraic topology. A more mathematical definition of the Betti numbers for a topological space X is given as the dimension of the kth homology group of X denoted as $H_k(X)$.

When analyzing data, one always has to extract as much information as possible. Three-dimensional configuration of structures often carries higher-order features and traditional data analysis tools tend to neglect these features. This is when topology comes into the picture (Ghrist, 2008). Algebraic topology has developed the necessary tools to properly treat and analyze these features. However, to analyze data using topology, one has to convert the data into a topological space. In the present situation, for instance, the geometric data describing a chemical structure has to be converted into a simplical complex. There are several ways one could do this. The approach we applied here builds the so-called Rips or Vietoris–Rips complex (Hausmann, 1995). This complex is relatively computationally cheap, compared to other complexes. When building a Rips complex from a set of points one has to define a distance parameter d. The complex is built by forming a simplex by every set of points which are closer to each other than d.

At this point, one may ask which value of d is the best for a given dataset. The answer is found in Ghrist (2008): Betti numbers for a single value for d is not enough, one needs to be able to decide also which holes are essential and which are only the result of fluctuations.

This idea gave rise to the concept of persistence. Persistence was introduced by Edelsbrunner, Letscher, and Zomorodian (2000). For a short motivation consider the $(R_i)_{i=\overline{1,N}}$ sequence of Rips complexes corresponding to the increasing sequence of $(d_i)_{i=\overline{1,N}}$ distance parameters. The following natural inclusion holds:

$$R_1 \xrightarrow{l} R_2 \xrightarrow{l} \cdots \xrightarrow{l} R_{N-1} \xrightarrow{l} R_N. \tag{A.7}$$

When analyzing persistence, instead of looking at individual terms of R_i one examines the iterated inclusions $l\colon H \times R_i \to H \times R_j$ when $i < j$ as the inclusion maps reveal which features persist. For more details and a very good overview see Ghrist (2008).

A.6. Hausdorff distance and the derived topological similarity

The concept of Hausdorff distance (or Pompeiu–Hausdorff distance) was developed in the field of set theory. Generally, the definition of the Hausdorff distance of the sets $C, D \subset \mathbb{R}^n$ is given by

$$d_H(C,D) = \sup_{x \in \mathbb{R}^n} |d_C(x) - d_D(x)|, \tag{A.8}$$

where $d_C(x)$ is the distance of the point x from the set C (Rockafellar & Wets, 1998). A more intuitive definition can be given as follows:

$$d_H(C,D) = \max\{\sup_{x \in C}\inf_{y \in D} d(x,y)\},\ \{\sup_{y \in D}\inf_{x \in C} d(y,x)\}. \tag{A.9}$$

In our case, we are dealing with subsets of $\mathbb{R}^2$ as each element—each bar—is characterized by two numbers. Thus, the Hausdorff distance can be applied straightforwardly.

We represent each of our objects with three different sets: persistence intervals for number of connected components, number of holes, and number of voids, respectively. It is not correct to compare the number of holes to the number of connected components. Therefore, we have to calculate the Hausdorff distance for the corresponding pair of sets. In a more formal description, let A and B be two objects for which the set representing the barcodes for connected components is denoted by CC_A and CC_B, the set representing the barcodes for holes is denoted by HL_A and HL_B, while the set representing the barcodes for voids is denoted by VD_A and VD_B, respectively. We chose to define our topological similarity of A and B as

$$\text{TSM} = 1 - [d_H(CC_A, CC_B) + d_H(HL_A, HL_B) + d_H(VD_A, VD_B)]. \quad \text{(A.10)}$$

REFERENCES

Bohn, M., & Heermann, D. W. (2010). Diffusion-driven looping provides a consistent framework for chromatin organization. *PLoS One, 5*(8), e12218.

Botta, M., Haider, S., Leung, I. X., Lio, P., & Mozziconacci, J. (2010). Intra- and inter-chromosomal interactions correlate with ctcf binding genome wide. *Molecular Systems Biology, 6*, 426.

Carlsson, G. (2009). Topology and Data. *Bulletin of the American Mathematical Society, 46*, 255–308.

Cremer, T., & Cremer, M. (2010). Chromosome territories. *Cold Spring Harbor Perspectives in Biology, 2*(3), a003889.

de Gennes, P. G. (1979). *Scaling concepts in polymer physics*. Ithaca, NY: Cornell University Press.

Dekker, J. (2003). A closer look at long-range chromosomal interactions. *Trends in Biochemical Sciences, 28*(6), 277–280.

de Laat, W., & Grosveld, F. (2003). Spatial organization of gene expression: The active chromatin hub. *Chromosome Research, 11*(5), 447–459.

Delisi, C. (1974). Statistical thermodynamics of oligomer–polymer interactions. *Biopolymers, 13*, 2305–2314.

De Lucchini, S., & Cardellini, P. (2000). The Xenopus laevis zinc finger protein gene Xfin: Developmental expression and in vivo functional studies. *Italian Journal of Zoology, 67*, 45–49.

de Pablo, J. J. (2011). Coarse-grained simulations of macromolecules: From DNA to nanocomposites. *Annual Review of Physical Chemistry, 62*, 555–574.

Edelsbrunner, H., Letscher, D., & Zomorodian, A. (2000). Topological persistence and simplification. In: *Proceedings 41st annual symposium on foundations of computer science, 2000* (pp. 454–463).

Elrod-Erickson, M., Rould, M. a., Nekludova, L., & Pabo, C. O. (1996). Zif268 protein-DNA complex refined at 1.6 A: A model system for understanding zinc finger-DNA interactions. *Structure (London, England: 1993), 4*(10), 1171–1180.

Feeney, A. J., & Verma-Gaur, J. (2012). CTCF-cohesin complex: Architect of chromatin structure regulates V(D)J rearrangement. *Cell Research, 22*(2), 280–282.

Filippova, G. N. (2008). Genetics and epigenetics of the multifunctional protein ctcf. *Current Topics in Developmental Biology, 80*, 337–360.

Fritsche, M., Laura, G., Reinholdt, M. L., Handel, M. A., Bewersdorf, J., & Heermann, D. W. (2012). The impact of entropy on the spatial organization of synaptonemal complexes within the cell nucleus. *PLoS One, 7*(5), e36282.

Ghrist, R. (2008). Barcodes: The persistent topology of data. *Bulletin of the American Mathematical Society, 45*, 61–75.

Guelen, L., Pagie, L., Brasset, E., Meuleman, W., Faza, M. B., Talhout, W., et al. (2008). Domain organization of human chromosomes revealed by mapping of nuclear lamina interactions. *Nature, 453*(7197), 948–951.

Hadjur, S., Williams, L. M., Ryan, N. K., Cobb, B. S., Sexton, T., Fraser, P., et al. (2009). Cohesins form chromosomal cis-interactions at the developmentally regulated IFNG locus. *Nature, 460*(7253), 410–413.

Hausmann, J.-C. (1995). On the Vietoris-Rips complexes and a cohomology theory for metric spaces. In *Annals of Mathematics Studies: Vol. 138. Prospects in topology (Princeton, NJ, 1994)* (pp. 175–188). Princeton, NJ: Princeton Univ. Press.

Heermann, D. W. (2011). Physical nuclear organization: Loops and entropy. *Current Opinion in Cell Biology, 23*(3), 332–337.

Heermann, D. W. (2012). Mitotic chromosome structure. *Experimental Cell Research, 318* (12), 1381–1385.

Heermann, D. W., & Binder, K. (2010). *Monte carlo simulation in statistical physics.* Heidelberg: Springer-Verlag.

Huang, N., Shoichet, B. K., & Irwin, J. J. (2006). Benchmarking sets for molecular docking. *Journal of Medicinal Chemistry, 49*(23), 6789–6801.

Hyre, D. E., & Klevit, R. E. (1998). A disorder-to-order transition coupled to DNA binding in the essential zinc-finger DNA-binding domain of yeast ADR1. *Journal of Molecular Biology, 279*(4), 929–943.

Jantz, D., & Berg, J. M. (2004). Reduction in DNA-binding affinity of Cys2His2 zinc finger proteins by linker phosphorylation. *Proceedings of the National Academy of Sciences of the United States of America, 101*(20), 7589–7593.

Jerabek, H., & Heermann, D. W. (2012). Expression-dependent folding of interphase chromatin. *PLoS One, 7*(5), e37525.

Kim, J. S., & Pabo, C. O. (1998). Getting a handhold on DNA: Design of polyzinc finger proteins with femtomolar dissociation constants. *Proceedings of the National Academy of Sciences of the United States of America, 95*(6), 2812–2817.

Kolomeisky, A. B. (2011). Physics of protein-DNA interactions: Mechanisms of facilitated target search. *Physical Chemistry Chemical Physics: PCCP, 13*(6), 2088–2095.

Kornberg, R. D., & Stryer, L. (1988). Statistical distributions of nucleosomes: nonrandom locations by a stochastic mechanism. *Nucleic Acids Research, 16*(14), 6677–6690.

Kurukuti, S., Tiwari, V. K., Tavoosidana, G., Pugacheva, E., Murrell, A., Zhao, Z., et al. (2006). Ctcf binding at the h19 imprinting control region mediates maternally inherited higher-order chromatin conformation to restrict enhancer access to igf2. *Proceedings of the National Academy of Sciences of the United States of America, 103*(28), 10684–10689.

Laity, J. H., Dyson, H. J., & Wright, P. E. (2000). DNA-induced alpha-helix capping in conserved linker sequences is a determinant of binding affinity in Cys(2)-His(2) zinc fingers. *Journal of Molecular Biology, 295*(4), 719–727.

Lieberman-Aiden, E., van Nynke, L., Berkum, L. W., Imakaev, M., Ragoczy, T., Telling, A., et al. (2009). Comprehensive mapping of long-range interactions reveals folding principles of the human genome. *Science, 326*(5950), 289–293.

Madras, N., & Sokal, A. D. (1988). The pivot algorithm: A highly efficient monte carlo method for the self-avoiding walk. *Journal of Statistical Physics, 50*(1), 109–186.

Mateos-Langerak, J., Bohn, M., de Leeuw, W., Giromus, O., Manders, E. M., Verschure, P. J., et al. (2009). Spatially confined folding of chromatin in the interphase nucleus. *Proceedings of the National Academy of Sciences of the United States of America, 106* (10), 3812–3817.

Mishiro, T., Ishihara, K., Hino, S., Tsutsumi, S., Aburatani, H., Shirahige, K., et al. (2009). Architectural roles of multiple chromatin insulators at the human apolipoprotein gene cluster. *The EMBO Journal, 28*(9), 1234–1245.

Nanda, V. (2012). Perseus: The persistent homology software. http://www.math.rutgers.edu/vidit/perseus.html. Accessed 15.10.12.

Nelson, D. L., & Cox, M. M. (2004). *Lehninger Principles of Biochemistry.* W. H. Freeman, ISBN-10:0716743396.

Ohlsson, R., Renkawitz, R., & Lobanenkov, V. (2001). CTCF is a uniquely versatile transcription regulator linked to epigenetics and disease. *Trends in Genetics: TIG, 17*(9), 520–527.

Pabo, C. O., Peisach, E., & Grant, R. A. (2001). Design and selection of novel Cys2His2 zinc finger proteins. *Annual Review of Biochemistry, 70*, 313–340.

Paillard, G., Deremble, C., & Lavery, R. (2004). Looking into DNA recognition: Zinc finger binding specificity. *Nucleic Acids Research, 32*(22), 6673–6682.

Pavletich, N. P., & Pabo, C. O. (1991). Zinc finger-DNA recognition: Crystal structure of a Zif268-DNA complex at 2.1 A. *Science, 252,* 809–817.

Phillips, J. E., & Corces, V. G. (2009). CTCF: Master weaver of the genome. *Cell, 137*(7), 1194–1211.

Poirier, M. G., Bussiek, M., Langowski, J., & Widom, J. (2008). Spontaneous access to DNA target sites in folded chromatin fibers. *Journal of Molecular Biology, 379*(4), 772–786.

Rebar, E. J., & Pabo, C. O. (1994). Zinc finger phage: Affinity selection of fingers with new DNA-binding specificities. *Science (New York, N.Y.), 263*(5147), 671–673.

Rockafellar, R. T., & Wets, R. J. B. (1998). Set convergence. In *Grundlehren der mathematischen Wissenschaften: Vol. 317. Variational analysis* (pp. 108–147). Berlin Heidelberg: Springer. http://dx.doi.org/10.1007/978-3-642-02431-34.

Rogers, D. J., & Tanimoto, T. T. (1960). A computer program for classifying plants. *Science, 132*(3434), 1115–1118.

Rubio, E. D., Reiss, D. J., Welcsh, P. L., Disteche, C. M., Filippova, G. N., Baliga, N. S., et al. (2008). Ctcf physically links cohesin to chromatin. *Proceedings of the National Academy of Sciences, 105*(24), 8309–8314.

Ryan, R. F., & Darby, M. K. (1998). The role of zinc finger linkers in p43 and TFIIIA binding to 5S rRNA and DNA. *Nucleic Acids Research, 26*(3), 703–709.

Sander, J. D., Zaback, P., Keith Joung, J., Voytas, D. F., & Dobbs, D. (2007). Zinc Finger Targeter (ZiFiT): An engineered zinc finger/target site design tool. *Nucleic Acids Research, 35*(suppl 2), W599–W605.

Segal, E., & Widom, J. (2009). From DNA sequence to transcriptional behaviour: A quantitative approach. *Nature Reviews. Genetics, 10*(7), 443–456.

Sexton, T., Bantignies, F., & Cavalli, G. (2009). Genomic interactions: Chromatin loops and gene meeting points in transcriptional regulation. *Seminars in Cell & Developmental Biology, 20*(7), 849–855.

Simonis, M., Klous, P., Splinter, E., Moshkin, Y., Willemsen, R., de Wit, E., et al. (2006). Nuclear organization of active and inactive chromatin domains uncovered by chromosome conformation capture-on-chip (4c). *Nature Genetics, 38*(11), 1348–1354.

Simonis, M., Kooren, J., & de Laat, W. (2007). An evaluation of 3c-based methods to capture dna interactions. *Nature Methods, 4*(11), 895–901.

Slutsky, M., & Mirny, L. a. (2004). Kinetics of protein-DNA interaction: Facilitated target location in sequence-dependent potential. *Biophysical Journal, 87*(6), 4021–4035.

Splinter, E., Heath, H., Kooren, J., Palstra, R.-J., Klous, P., Grosveld, F., et al. (2006). Ctcf mediates long-range chromatin looping and local histone modification in the beta-globin locus. *Genes & Development, 20*(17), 2349–2354.

Verschure, P. J., van Der Kraan, I., Manders, E. M., & van Driel, R. (1999). Spatial relationship between transcription sites and chromosome territories. *The Journal of Cell Biology, 147*(1), 13–24.

Vuzman, D., Polonsky, M., & Levy, Y. (2010). Facilitated DNA search by multidomain transcription factors: Cross talk via a flexible linker. *Journal of Biophysics, 99*(4), 1202–1211.

Wendt, K. S., & Peters, J.-M. (2009). How cohesin and ctcf cooperate in regulating gene expression. *Chromosome Research, 17*(2), 201–214.

Willett, P., Barnard, J. M., & Downs, G. M. (1998). Chemical similarity searching. *Journal of Chemical Information and Computer Sciences, 38*(6), 983–996.

Winter, R. B., Berg, O. G., & Von Hippel, P. H. (1981). Diffusion-driven mechanisms of protein translocation on nucleic acids 3. The Escherichia coli lac repressor-operator interaction: Kinetic measurements and conclusions. *Biochemistry, 20,* 6961–6977.

Wolfe, S. A., Nekludova, L., & Pabo, C. O. (2000). DNA recognition by Cys2His2 zinc finger, proteins. *Annual review of biophysics and biomolecular structure, 29,* 183–212.

Zandarashvili, L., Vuzman, D., Esadze, A., Takayama, Y., Sahu, D., Levy, Y., et al. (2012). Asymmetrical roles of zinc fingers in dynamic DNA-scanning process by the inducible transcription factor Egr-1. *Proceedings of the National Academy of Sciences of the United States of America, 109*, E1724–E1732.

Zhang, Y., & Heermann, D. W. (2011). Loops determine the mechanical properties of mitotic chromosomes. *PLoS One, 6*(12), e29225.

Zhou, H.-x. (2001). The Affinity-Enhancing Roles of Flexible Linkers in Two-Domain DNA-Binding. *Biochemistry, 40*, 15069–15073.

Zlatanova, J., & Caiafa, P. (2009). Ctcf and its protein partners: Divide and rule? *Journal of Cell Science, 122*(Pt 9), 1275–1284.

CHAPTER FOUR

Dynamics of Modeled Oligonucleosomes and the Role of Histone Variant Proteins in Nucleosome Organization

Amutha Ramaswamy[*,1], Ilya Ioshikhes[†]

[*]Centre for Bioinformatics, School of Life Sciences, Pondicherry University, Kalapet, Pondicherry, India
[†]Ottawa Institute of Systems Biology (OISB), Department of Biochemistry, Microbiology and Immunology (BMI), University of Ottawa, Ottawa, Ontario, Canada
[1]Corresponding author: e-mail address: amutha_ramu@yahoo.com

Contents

Abstract

Elucidation of the structural dynamics of a nucleosome is of primary importance for understanding the molecular mechanisms that control the nucleosomal positioning. The presence of variant histone proteins in the nucleosome core raises the functional diversity of the nucleosomes in gene regulation and has the profound epigenetic consequences of great importance for understanding the fundamental issues like the assembly of variant nucleosomes, chromatin remodeling, histone posttranslational modifications, etc. Here, we report our observation of the dominant mechanisms of relaxation motions of the oligonucleosomes such as dimer, trimer, and tetramer (in the beads on a string model) with conventional core histones and role of variant histone

Advances in Protein Chemistry and Structural Biology, Volume 90
ISSN 1876-1623
http://dx.doi.org/10.1016/B978-0-12-410523-2.00004-3

H2A.Z in the chromatin dynamics using normal mode analysis. Analysis of the direction-ality of the global dynamics of the oligonucleosome reveals (i) the in-planar stretching as well as out-of-planar bending motions as the relaxation mechanisms of the oligonucleosome and (ii) the freedom of the individual nucleosome in expressing the combination of the above-mentioned motions as the global mode of dynamics. The highly dynamic N-termini of H3 and (H2A.Z–H2B) dimer evidence their participation in the transcriptionally active state. The key role of variant H2A.Z histone as a major source of vibrant motions via weaker intra- and intermolecular correlations is empha-sized in this chapter.

1. INTRODUCTION

Nucleosome, the fundamental repeating structural/functional unit of chromatin, plays a central role in chromatin biology and hence in the gene regulation as well (Van Holde, 1989; Wolffe, 1998). In the nucleosome, the genetic DNA of length about 147 base pairs is wrapped tightly around a dyad symmetrical complex of eight histone protein molecules of four different kinds (two copies of H2A, H2B, H3, and H4 histones). The nucleosomes occupy ~75% of the eukaryotic genomes with an average distance of ~200 bp between their centers (though this distance may vary for different species and genome segments) and represent a basic level of organization of chromatin (DNA–protein complex in eukaryotic cell nuclei). The DNA segment between two adjacent nucleosomes is named linker. The canonical nucleosome structure is stabilized with DNA through the linker histone, H1 (or H5) (McGhee & Felsenfeld, 1980). Organization of these histone pro-teins has major impact on regulating all cellular processes like transcription, replication, recombination, and repair. The dynamics of the nucleosome is highly dependent on the sequence and structure of the histone proteins, which affects the regulation of gene expression and other genetic processes in chromatin (Zlatanova, Bishop, Victor, Jackson, & van Holde, 2009). The detailed nucleosome structure, the general organization of the chromatin, and the mechanism of transcriptional regulation as result of the nucleosome dynamics have been the issue of active investigation over many years (Ramakrishnan, 1997; Workman & Kingston, 1998). The structural prop-erties of nucleosomes and higher order chromatin organization are highly modulated by a variety of histone posttranslational modifications such as acetylation, phosphorylation, and methylation (Kouzarides, 2007). In addi-tion to these histone modifications, the noncanonical histone variants that differ at the amino acid sequence level (from a few positions to large protein domains) express distinct nucleosomal architecture and heterogeneity (Zweidler, 1984) and hence play major roles in nuclear processes such as DNA repair, chromosome segregation, transcription initiation and termination, sex chromosome condensation, sperm chromatin packaging, etc. In this context, understanding the mode of incorporation of histone variants in the nucleosome, the structure of the variant chromatin, and posttranslational modifications that occur on the variants is of great importance. This chapter reveals important aspects of the chromatin dynamics based on the comparative analyses of the nucleosomes containing the canonical histone proteins and variants thereof.

1.1. The structure of nucleosome

In eukaryotes, the bulky linear DNA is packed by histone proteins such as H3, H4, H2A, and H2B to fit in a small compartment of the cell nucleus. The DNA is also required for the stability of histone octamer, and in its absence the histone octamer dissociates into two dimers of H2A–H2B and a tetramer of H3–H4 in physiological salt concentrations (Chung, Hill, & Doty, 1978). The transcriptional and structural status of this DNA is highly influenced by the degree of packaging and its specific mechanisms such as DNA bending, electrostatic interactions between DNA and the histones. The full nascent nucleosome (released by the initial attack of micrococcal nuclease) contains 160–240 base pairs of DNA in an unstable state (Kornberg, 1977) and is partially stabilized when the nuclease begins to digest the terminal DNA originally derived from the linker region between adjacent nucleosomes (Axel, 1975; Greil, Igo-Kemenes, & Zachau, 1976; Noll & Kornberg, 1977; Shaw, Herman, Kovacic, Beaudreau, & Van Holde, 1976; Simpson & Whitlock, 1976; Sollner-Webb & Felsenfeld, 1975; Simpson, 1978). Among these are particles with 160–170 base pairs of DNA bound to histone H1 (Noll & Kornberg, 1977; Varshavsky et al., 1976). Further digestion produces a more stable nucleosome particle containing about 146 base pairs of DNA and an octamer consisting of two copies of each of the four core histones (H2A, H2B, H3, and H4) with no interacting H1 (Bryan, Wright, & Olins, 1979; Lutter, 1979; Prunell et al., 1979; Simpson & Kunzler, 1979). A single nucleosome contains 14 noncovalent histone–DNA contacts (Luger, Mader, Richmond, Sargent, & Richmond, 1997). The histone proteins are highly positively charged (as having amino acid sequences rich in lysine or arginine) and hence attract the negatively charged phosphate groups of the double helix backbone during DNA condensation. The histone–DNA binding is also affected by DNA deformational properties and hence is DNA sequence specific, that is, some DNA segments have higher affinity to the histone octamer binding than others. Intruding these sites by chromatin remodeling factors result in nucleosome repositioning, disassembly, etc. (Cairns, 2009; Clapier & Cairns, 2009). The secondary structure of the chromatin is determined by the linear organization of nucleosomes with the linker DNA which determines the spatial orientation of successive nucleosomes in a folded array and distance between adjoining nucleosomes (Arya, Maitra, & Grigoryev, 2010; Grigoryev, Arya, Correll, Woodcock, & Schlick, 2009). The histone proteins are highly conserved across all eukaryotes and so is the basic nucleosome structure.

1.2. The histone octamer core

The histone octamer nucleosome core, formed by two copies of H2A, H2B, H3, and H4 histones, is wrapped by the DNA of length $\sim$147 bp in a left-handed superhelix fashion (Luger et al., 1997; Luger et al., 1999a; Luger, Rechsteiner, & Richmond, 1999b; Richmond, Finch, Rushton, Rhodes, & Klug, 1984). The assembly of a stable nucleosome core depends on the initial heterodimerization of H3 with H4 with the subsequent dimerization to form the $(H3–H4)_2$ tetramer (Eickbush & Moudrianakis, 1978), followed by the dimerization of histones H2A and H2B that bind to both sides of the $(H3–H4)_2$ tetramer (Fig. 4.1) (Hayes, Clark, & Wolffe, 1991; Hayes, Tullius, & Wolffe, 1990). The interactions between the octameric histone proteins are divided into two categories: strong interactions between the pairs H2A–H2B, H2B–H4, and H3–H4 and weak interactions between H2A–H3, H2B–H3, and H2A–H4 pairs (Bonner & Pollard, 1975; Isenberg, 1979). The nucleosome cores are separated by the linker DNA of variable length, and the linker histone H1 (or H5) stabilizes the chromatin structure forming a bigger histone–DNA complex named chromatosome.

1.3. Histones

The core histones are characterized by the presence of a structurally conserved motif called histone-fold domain and the dynamic histone-fold extensions, named histone tails, which highly regulate the nucleosome stability (Zheng & Hayes, 2003) and hence play a major role in defining the chromatin higher order structures (Dorigo, Schalch, Bystricky, & Richmond, 2003). The histone fold consists of three α-helices (α1, α2, and α3) connected by short loops L1 and L2 that mediate heterodimeric

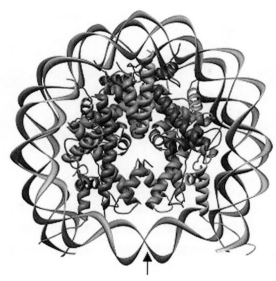

Figure 4.1 Architecture of the octameric histones H3 (pink), H4 (green), H2A.Z (blue), and H2B (orange) and DNA (gray) in the nucleosome with the histone variant H2A.Z (Suto, Clarkson, Tremethick, & Luger, 2000). Dark and light colors distinguish each copy of the monomers. Arrow indicates the dyad axis. (For interpretation of the references to color in this figure legend, the reader is referred to the online version of this chapter.)

interactions between the core histones (Arents et al., 1991). In the process of the dimerization, the loop L1 of one histone aligns with the loop L2 of the other forming so named handshake motif (Fig. 4.2). This handshake motif interacts with DNA via two L1L2 sites and one $\alpha 1\alpha 1$ site. The flexible tails of the core histones interact with DNA via the minor groove. The histone tail domains are the major targets for posttranslational modifications such as acetylation, methylation, and phosphorylation and so are the key arbiters of chromatin gene regulatory function (Davie, 1998).

1.4. Histone variants

The functional state of the chromatin is highly determined by the structural and functional roles of the canonical histone proteins involved in all stages of chromatin formation from the protein–DNA complexation to genome packaging and regulation (Zlatanova et al., 2009). Histone proteins have a high degree of conservation due to their role in maintaining the nucleosome octamer core. The existence of noncanonical histone proteins (histone variants) creates puzzles on the assembly of variant nucleosomes, chromatin

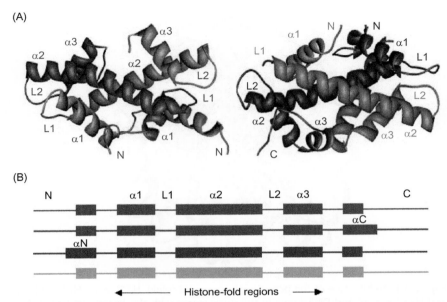

Figure 4.2 (A) The secondary structure of handshake motifs; H3–H4 and H2A–H2B formed by the histones H3 (pink), H4 (green), H2A.Z (blue), and H2B (orange) and (B) their schematic representation as colored in (A). (See Color Insert.)

remodeling, and histone posttranslational modification, etc. The interplay between the variant histones results in a highly dynamic chromosome architecture, and the profound epigenetic consequences are of great importance for gene regulation. Understanding the influence of the variant histones over the structure and dynamics of chromatin would reveal the fundamental mechanisms involved in functions such as nucleation of variant histone folds and the assembly of histone variants into chromatin, the structure of the variant chromatin, posttranslational modifications that occur on the variants, etc. Hence, recent studies focus on the analysis of chromatin structure with variant histones. Some of the well-known eukaryotic histone variants are listed in Table 4.1.

Natural types of histones occur in the form of various isoforms (H2A.1, H2A.2), variants (H2A.Z, H2A.X, H3.3, and CENP-A), and histone-like proteins (macroH2A). The H3-like protein, centromere protein A (CENP-A), found at mammalian centromeres is the first histone variant (Palmer, O'Day, Trong, Charbonneau, & Margolis, 1991). The H3-like counterparts of CENP-A in all eukaryotic lineages are collectively called CenH3s (Malik & Henikoff, 2003). The other eukaryotic H3 "replacement" variant, H3.3

Table 4.1 List of eukaryotic histone variants

Family	Variant	Function	References
H1	Canonical		Thomas (1999)
	$H1^0$	Transcription repression	Doenecke, Albig, Bouterfa, and Drabent (1994)
	H5	Transcription repression	Zhou, Gerchman, Ramakrishnan, Travers, and Muyldermans (1998)
H2A	Canonical	Genome packaging	Eickbush, Watson, and Moudrianakis (1976)
	H2A.X	DNA ds break repair	Celeste et al. (2003)
	H2A.Z	Promoter insulation	Jackson, Falciano, and Gorovsky (1996)
	MacroH2A	X chromosome inactivation	Pehrson and Fuji (1998)
	MacroH2A1	promote interactions between nucleosomes	Costanzi and Pehrson (1998)
	MacroH2A1.2	meiotic sex chromosome inactivation	Changolkar and Pehrson (2002)
	H2A.Bbd	Transcription activation	Bao et al. (2004)
H2B	Canonical	Genome packaging	Sivolob, Lavelle, and Prunell (2003)
	TH2B	Testes specific	Shires, Carpenter, and Chalkley (1976)
	H2BFWT	Testes specific	Churikov et al. (2004)
	H2BV	Transcription initiation	Lowell, Kaiser, Janzen, and Cross (2005)
H3	Canonical	Genome packaging	Garcia et al. (2005)
	H3.1	Genome packaging	Tagami, Ray-Gallet, Almouzni, and Nakatani (2004)
	H3.2	Genome packaging	Garcia et al. (2005)
	H3.3	Genome packaging/ replacement	Tagami et al. (2004)

Continued

Table 4.1 List of eukaryotic histone variants—cont'd

Family	Variant	Function	References
	H3.4	Genome packaging	Ahmad and Henikoff (2002a)
	CenH3	Centromere identity	Palmer, O'Day, Wener, Andrews, and Margolis (1987)
	H3t	Testes specific	Tachiwana, Osakabe, Kimura, and Kurumizaka (2008)
	H3.X, H3.Y	Regulation of cellular responses to outside stimuli	Wiedemann et al. (2010)
H4	Canonical	Genome packaging	Sierra, Stein, and Stein (1983)

differs from H3 at only four amino acid positions and is actively involved throughout the cell cycle (Ahmad & Henikoff, 2002a,2002b; Hendzel & Davie, 1990). Another important histone variant associated with active chromatin is H2A.Z (Jackson & Gorovsky, 2000; Malik & Henikoff, 2003). H2A.Z is essential for the viability of many organisms and has functions distinct from those of the major H2A histone in chromatin. Several pioneers have reported the importance and the functional diversity of the nucleosome by H2A.Z (Abbott, Ivanova, Wang, Bonner, & Ausio, 2001; Adam, Robert, Larochelle, & Gaudreau, 2001; Brown, 2001; Clarkson, Wells, Gibson, Saint, & Tremethick, 1999; Fan, Gordon, Luger, Hansen, & Tremethick, 2002; Jackson & Gorovsky, 2000; Suto et al., 2000).

The crystal structures of nucleosome containing both H2A and H2A.Z show significant structural similarities (Suto et al., 2000). Despite the structural similarity, the nucleosome containing H2A.Z is significantly less stable when compared to the nucleosome with H2A and hence is involved in transcriptional activation (Fan et al., 2002). Other H2A variants such as MacroH2A and H2A-Bbd are associated in distinguishing the inactive X chromosome from the active one, respectively (Chadwick & Willard, 2001, 2002). Thus, histone variants express precedence in determining the specification and inheritance of chromatin states.

The functional heterogeneity of the histone variants has been well documented by Brown (2001). Characterization of H2A.Z nucleosome crystal structure revealed a snapshot of histone–DNA and histone–histone interactions within the nucleosome core particle containing the histone variant (Davie, 1998). The substitution of Gln104 (in H2A) by Gly (in H2A.Z) destabilizes the association of 2(H2A.Z–H2B)–(H3–H4)$_2$ by loosing several hydrogen bonds and promotes DNA activation (Clarkson et al., 1999; Santisteban, Kalashnikova, & Smith, 2000). The γH2AX, which is the phosphorylated H2AX variant at serine 139, has been widely used as a sensitive marker for DNA double-strand breaks (Yuan, Adamski, & Chen, 2010).

1.5. Higher order structure of chromatin

The array of nucleosomes in a "beads on a string" model is stabilized by the binding of H1 (or H5) through its globular domain and seals the two full turns of DNA (~166 bp) around the octamer (Thomas, 1999). The globular domain of H1 (or H5), which is a three-helix bundle showing structural homology to helix-turn-helix DNA-binding proteins, binds simultaneously to both the DNA duplexes of the nucleosome (Draves, Lowary, & Widom, 1992; Thomas, Rees, & Finch, 1992). Nucleosomes, connected by about 20–60 bp of linker DNA, form a 10-nm "beads on a string" array for its further condensed/compact structure of a 30-nm chromatin fiber (Horn & Peterson, 2002; Mohd-Sarip & Verrijzer, 2004; Richmond & Widom, 2000). Here, in this chapter, the oligonucleosomes are modeled like a beads-on-string model.

The organization of nucleosomes into the compact chromatin fiber is of great debate due to the structural complexity. Several models have been proposed for the chromatin fiber, and the most enduring models are (i) the two-start helical ribbon model (Woodcock, Frado, & Rattner, 1984; Worcel, Strogatz, & Riley, 1981), (ii) the two-start crossed-linker model (Williams et al., 1986), and (iii) the one-start solenoid model (Finch & Klug, 1976; Thoma, Koller, & Klug, 1979; Van Holde, Sahasrabuddhe, & Shaw, 1974; Widom & Klug, 1985). The architecture of the two-start helix consists of repeating units of nucleosomes folded into an irregular 3D zigzag arrangement (Horowitz, Koster, Walz, & Woodcock, 1997). In this model, the helix stacks with the nonsequential nucleosomes across from one another $(i+2)$ and twists to form a two stacks of winding nucleosomes in a superhelix (Williams et al., 1986). The two-start helical ribbon and the two-start crossed-linker differ by the orientation of the zigzag pattern with respect to the fiber axis. The two-start helical ribbon forms a parallel-zigzag arrangement (Woodcock et al., 1984; Worcel et al., 1981), and the two-start crossed-linker forms a perpendicular-zigzag arrangement (Williams et al., 1986). The zigzag orientation of the nucleosomal arrays is evidenced by many experimental reports. The crystal structure of a tetranucleosome evidences a two-start type of fiber (Schalch, Duda,

Sargent, & Richmond, 2005). The solenoid fiber model adopts a hand-to-hand orientation by the consecutive nucleosomes, and the nucleosomal chains coil around an inner cavity with six to eight nucleosomes per turn with a pitch of ~11 nm to form a one-start solenoid superhelix (Finch & Klug, 1976; McGhee, Nickol, Felsenfeld, & Rau, 1983; Thoma et al., 1979), in which nucleosome-stacking interactions occur between nearest neighbor nucleosomes $(i + 1)$.

To model this oligonucleosomes, a complete nucleosome structure with linker DNA is modeled using (i) the monomeric nucleosome structure (1F66/1EQZ) used for the global dynamics, (ii) a B–DNA of sequence 5′-CTGCAGATTCTACCAAAAG-3′ as the linker DNA region (after minimization) to facilitate the interaction with linker histone, and (iii) the structure of linker histone H1 from the Protein Data Bank (PDB): 1GHC (Cerf et al., 1994). The deformed DNA duplex terminus of the nucleosome 1F66/1EQZ was removed, and a new DNA fragments of same sequence (10 base pairs) were used after the customary minimization procedure to ensure both perfect base pair interactions (stacking and H–bonding) and connection between this new fragment and the linker DNA. A detailed description on the modeling of the nucleosome array in the beads–on–string approximation is well documented in Ramaswamy and Ioshikhes (2007).

1.6. Coarse grained dynamics

Molecular dynamics simulation is currently emerging as a competitive alternative approach in understanding the functional dynamics of biomolecules beyond the experimental limit. This atomistic simulation explicitly accounts every atom of the system as a point mass whereas the interatomic interactions are governed by the molecular mechanics force fields in which the bonded interactions (like bonds, angles, planar, and nonplanar torsion angles) are modeled using the harmonic potentials, and the nonbonded interactions are taken care of by the Lennard-Jones and Coulomb potentials (Leach, 2001). Applications of this atomistic simulation are successful for many biomolecular systems including nucleic acids, proteins, enzymes, and other macromolecules (Schlick, Collepardo-Guevara, Halvorsen, Jung, & Xiao, 2011). The main drawback of this approach is the computational time, which linearly varies with the square of the total number of interaction sites present in the system and finds limitations in macromolecular simulation due to the number of degrees of freedom. At this juncture, the coarse–grain modeling finds merit in handling large biomolecular system as it involves in grouping atoms into a single super-sites to reduce the number of interactions/degrees of freedom (Muller, Katsov, & Schick, 2006; Nielsen, Lopez, Srinivas, & Klein, 2004; Shillcock & Lipowsky, 2006; Venturoli, Sperotto, Kranenburg, & Smit, 2006).

In line to this, normal modes analysis (derived based on the elastic network formalism) has emerged as an efficient and physically meaningful tool to study the functional dynamics of macromolecular structures and assemblies. Bahar and coworkers have introduced the Gaussian Network Model (GNM) and its advanced version the Anisotropic Network Model (ANM) to understand the functional dynamics of the macromolecules (Atilgan et al., 2001; Bahar, Atilgan, & Erman, 1997; Haliloglu, Bahar, & Erman, 1997). These models consider the biomolecule as an elastic network (EN) and generate a connectivity matrix of the C atoms (the nodes), which is determined by the cut-off (r_c) distance for the pairs of interacting (via springs) amino acids. The topology of this network is represented by a connectivity (Kirchhoff) matrix whose eigenvalue decomposition yields the normal modes of motion near the equilibrium structure and the lowest frequency (and largest amplitude) modes of motions refer to the global dynamics. In this chapter, the global dynamics of oligonucleosomes with canonical and noncanonical histones are analyzed using these EN models to understand the effect of variant histones on the functional motions of oligonucleosomes. The generation of EN models for the oligonucleosome is detailed in Ramaswamy and Ioshikhes (2007).

2. ROLE OF H2A.Z IN THE DYNAMICS OF MODELED OLIGONUCLEOSOMES

2.1. Nucleosome monomer

As our first attempt, we studied the structural dynamics of nucleosome with regular histones and the role of variant histone H2A.Z and other histone isomers such as H2A.1 and H2B.2 in nucleosome dynamics. For this analysis, we have used the crystal structures 1EQZ, 1F66, and 1KX4 from PDB (Harp, Hanson, Timm, & Bunick, 2000; Schalch et al., 2005; Suto et al., 2000). The structure of 1F66 (formed by 769 amino acids) has the recombinant mouse H2A.Z and recombinant *Xenopus laevis* histones H2B, H3, and H4. The octameric histone core of 1KX4 is of *X. laevis* origin, and the H2A.1 sequence (in 1KX4) is closely similar to H2A sequence of *Gallus gallus* (1EQZ), whereas H2B.2 (in 1KX4) is identical to the H2B in *X. laevis* (1F66). The nucleosome of 1EQZ (formed by 883 amino acids) is from chicken (*G. gallus*) histone octamer. All these histone proteins expressed more than 95% sequence similarity except H2A.Z. The variant H2A.Z (in 1F66) shows considerable variation with H2A (in 1EQZ) and hence is expected to express localized changes when interacting with H2B and $(H3-H4)_2$.

The normal modes extracted using EN models (GNM or ANM), with cut-off distances 10, 15, and 18 Å for protein–protein, protein–DNA, and DNA–DNA interactions, respectively, describe the global dynamics of a nucleosome. The individual modes of motions contributing to the global dynamics were dissected to identify the slowest mode (i.e., global mode) that characterizes the overall dynamics of the nucleosome. The nucleosome with regular histone proteins (canonical nucleosome) expresses several hydrogen bonds between the histone proteins and DNA super helix. The histone–DNA interacting sites (belonging to L1, L2, and α1 regions) are (i) T45, K64, F84, and V117 of H3; (ii) R36, G48, H75, and E90 of H4; (iii) F27, V46, K79, Q104, and G106 of H2A; and (iv) S33, S53, and S84 of H2B; and their motions are restricted due to the presence of these hydrogen bonds. The presence of H2A.Z (in place of H2A) releases most of the observed H-bonds and expresses comparatively higher dynamics. Apart from these H-bonds, the residues Arg24-Ile30 of H4 involved in gene silencing (Johnson, Fisher-Adams, & Grunstein, 1992) express their active participation in functional dynamics.

Figure 4.3 depicts the intra- and interatomic correlations observed between H2A–H2B and H2A.Z–H2B dimers. The comparison of intra- and intermolecular correlations existing between the major H2A and the variant H2A.Z indicates significantly different patterns. The C-terminal of H2A.Z contributes majorly to the nucleosome stabilization as its docking domain interacts with several histone–histone and histone–DNA sites in the nucleosome including (i) the DNA entry/exit site, (ii) $(H3-H4)_2$-tetramer, and (iii) H2A–H2B. In 1F66, the H2A.Z residues Arg81-Lys119 located at the interface between the $(H3-H4)_2$ tetramer and the (H2A–H2B) dimer exhibit substantial decreases in their couplings to the helix–loop α1L1 on the same monomer (H2A.Z) and to the loop–helix L2α3 on the neighboring (H2B) monomer (see the portions of the map enclosed in the black boxes). The loss of these long-range correlations implies an inept cooperativity of motion in the nucleosome and so disruption of the correlated motions acts as a key factor in the "destabilization" of chromatin function by the enhanced mobility of H2A.Z. These observations are in accord with the experimentally observed chromatin destabilizing role of H2A.Z (Fan et al., 2002). The presence of H2A.Z docking domain also relocates the

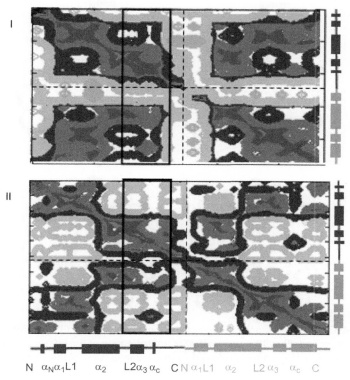

N $\alpha_N\alpha_1$L1 α_2 L2$\alpha_3\alpha_c$ C N α_1L1 α_2 L2 α_3 α_c C

Figure 4.3 The correlational map of the motions between the residues of H2A/H2A.Z and H2B in 1EQZ (I) and 1F66 (II) crystal structures. The uncorrelated residues (colored purple) separate the correlated (where amplitude increases from light green, dark green, and dark gray) and anticorrelated domain (colored gray) regions. (For interpretation of the references to color in this figure legend, the reader is referred to the online version of this chapter.)

αN helix of H3 (Suto et al., 2000) that interacts with the wound DNA and hence triggers the pattern of histone–DNA interactions.

The global dynamics of nucleosome resulted from the intra- and intermolecular motions revealed a highly symmetrical dynamics of nucleosome with respect to the dyad axis. The first two slow modes of the nucleosome dynamics are shown in a color-coded fashion [from black (rigid) to red (most flexible)] for the first (A) and second (B) slowest modes of structures 1EQZ and 1F66. In the first mode (Fig. 4.4) of the nucleosomes with and without variant histones, the rigid domains fall along the dyad axis of the nucleosome (locus of motion). The spatially conserved domains are identified at (1) the residues from L2 and C-terminus of H2B, (2) the loop L1 of H2A, (3) the residues from helix α3 to

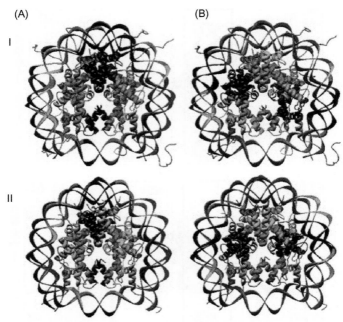

Figure 4.4 Color-coded representation of the dynamics of 1EQZ (I) and 1F66 (II) in the first (A) and second (B) collective modes. (For interpretation of the references to color in this figure legend, the reader is referred to the online version of this chapter.)

the C-tail of H3, and (4) the neighboring residues of His75 in H4. The N-termini of histones that are experimentally proven to regulate the eukaryotic transcription via protein–DNA interactions are highly dynamic in both modes. The dynamic N-tail of H3 is also engaged in a highly cooperative motion with the neighboring DNA segments. As a result, the wrapped DNA also exhibits a symmetric dynamics with respect to the dyad.

The locus of the second slowest mode of nucleosome differs between the H2A and H2A.Z nucleosomes. In the second slowest mode (Fig. 4.4) of the variant nucleosome, the central region of the nucleosome perpendicular to the dyad axis is highly rigid and forms the stable locus, while in the canonical nucleosome, the regions which are diagonal to the dyad axis form the locus of motion. The experimentally reported disordered N-termini of the histone proteins also express high mobilities. In addition, the DNA strand interacting sites of the histone proteins like (i) loop L1 and C-terminus of H2A, (ii) helix $\alpha 2$ and C-terminus of H2B, (iii) loop L2 and N-tail domain of H3, and (iv) L1 loop of H4 show the highest mobilities and influence the dynamics of DNA too.

It is interesting to correlate the experimental report on the transcriptionally active conformation of the nucleosome having variant H2A.Z due to the perturbed dynamics of both (H2A.Z–H2B) dimers and N-terminus of H3 in physiological conditions to the observed second slowest mode that expresses significantly different pattern of dynamics by the presence of H2A.Z histone, as the first mode is invariably preserved in both the structures.

The direction of the nucleosome motions studied using the deformation vector, Δr (which represents the extent of atomic fluctuations corresponding to the global mode of all residues and nucleotides), derived from ANM also revealed active conformational dynamics. Figure 4.5 shows the conformations generated by adding and subtracting Δr to–from the crystal structure coordinates (gray). Conformations at different views are displayed and the directions of deformations are shown by arrows for an easy understanding. Figure 4.5A shows the front view of the nucleosome and B illustrates the side view along the dyad axis.

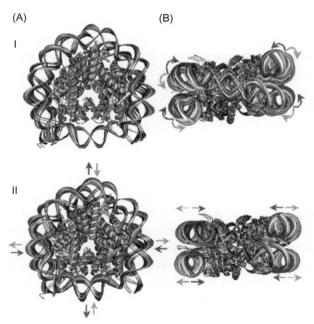

Figure 4.5 Superposition of the crystal structure (gray) and the two fluctuating conformations of 1F66 found by ANM analysis, by adding (magenta) and subtracting (cyan) the residue displacements driven in modes 1 (I) and 2 (II). (A) The front view and (B) the view along the dyad axis. (For interpretation of the references to color in this figure legend, the reader is referred to the online version of this chapter.)

In the first mode (Fig. 4.5I), the regions along the dyad are considerably rigid, while the other two sides of the nucleosome are fluctuating perpendicularly to the plane of dyad and hence result in the bending motion of the nucleosome in an out-of-plane fashion with respect to the dyad axis. Such bending motions would favor the dissociation of DNA from the histone octamer as evidenced by the experimental studies at relatively low ionic strength. In the second slowest mode of 1F66 (Fig. 4.5II), the nucleosome tends to compress and relax with respect to the dyad axis, and the vibrations are constrained to the plane of the nucleosome. Overall, the nucleosome expresses a breathing motion along the dyad axis, with a massive distortion of nucleosomal DNA mediated by histone–DNA interactions. The nucleotide interacting residues like (i) $\alpha1$ and L1 of H2A, (ii) L2 of H2B, (iii) L1 and L2 of H3, and (iv) L2 of H4 are heavily constrained in the canonical nucleosome. The interacting nucleotides of DNA undergo relatively larger amplitude motions in the nucleosome with H2A.Z, which support the weakening of the intermolecular interactions as reported experimentally. Hence, the interactions between the histone octamer and the wound DNA are instrumental in regulating the dynamics of the octamer, and conversely, the closely interacting proteins significantly alter the dynamics of the DNA, signaling the important role of DNA complexation with histones in regulating the transcriptional activity.

2.2. Nucleosome dimer

The dynamics of nucleosome complexes (oligonucleosomes) modeled using canonical and noncanonical (variant H2A.Z) histones was studied based on the structural dynamics of the single nucleosome (Ramaswamy, Bahar, & Ioshikhes, 2005). The dynamics of the oligonucleosomes such as dimer, trimer, and tetramer was studied using EN models (GNM and ANM) with larger cut-off distances to account for the complete protein–DNA interactions. The methodology adopted to model these oligonucleosomes is well documented in Ramaswamy and Ioshikhes (2007). The nucleosome dimer was modeled based on the experimentally resolved tetrameric conformation of a nucleosome reported by Schalch et al. (2005). The structure of the modeled dimer was compared to that of the nucleosome tetramer to ensure the existence of similar inter- and intradomain interactions as reported in the experimentally observed dimer conformation. Figure 4.6 shows the pattern of correlated and anticorrelated domain motions of the dimers H3–H4 (A) and H2A–H2B (B) in crystal structures 1ZBB (Fig. 4.6I), 1EQZ (Fig. 4.6II), and 1F66 (Fig. 4.6III) and reveals (i) the robustness of the modeled dimer

(A)   (B)

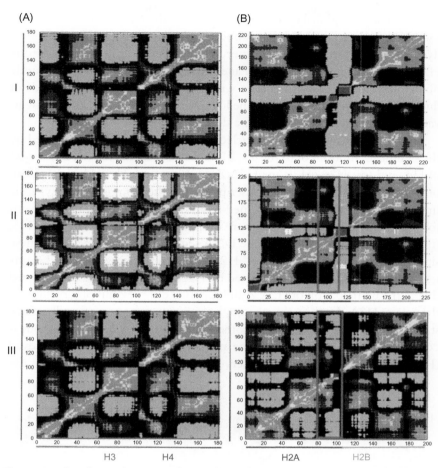

                        H3   H4   H2A   H2B

Figure 4.6 Correlational motions observed between the residues of dimers, H3–H4 (A) and H2A–H2B (B) in the first nucleosomes of 1ZBB (I), 1EQZ (II), and 1F66 (III) crystal structures. The uncorrelated regions (colored black) separate the correlated (where the amplitude increases from blue to red) and anti-correlated regions (colored cyan). (For interpretation of the references to color in this figure legend, the reader is referred to the online version of this chapter.)

through the correlational motions and (ii) the distinguished correlation pattern due to the presence of variant histone H2A.Z in place of H2A. It is obvious that the correlations between the motions of the handshake H2A–H2B (or H2A.Z–H2B) motifs of the nucleosome dimer are similar to the correlation of H2A–H2B (or H2A.Z–H2B) motifs of a single nucleosome and are also the same for H3–H4 and hence this analysis was not extended to the variant nucleosomes trimer and tetramer.

The first three global modes (A, B, and C) of the nucleosome dimer with canonical and variant H2A.Z histone are shown in Fig. 4.7. The global dynamics of the nucleosome dimer with canonical and variant histone H2A.Z is shown in panel I and II. Panel III depicts the side view of the mode observed in the nucleosome dimer with H2A.Z variant. The amplitude of fluctuation is graded from black (the lowest) to red (the highest) to show its variation from the rigid to dynamic domain.

The global dynamics of the nucleosome dimer is highly determined by the significant motions of (i) the $\alpha 1$, L1, and $\alpha 2$ of H3; (ii) the $\alpha 2$, L2, and $\alpha 3$ of H4; and (iii) H2A–H2B (or H2A.Z–H2B) dimer. The H3 N-terminal, one of the important sites for transcription modification, is highly dynamic in this mode. The highly constrained regions that act as the hinge domains for the free dynamics of the dimer are (i) linker DNA; (ii) the L2, $\alpha 3$ of H3; (iii) the L1, $\alpha 1$ of H4; and (iv) the DNA segments that interact with these sites. Even though the dimer H2A–H2B completely involves in this dynamics, the degree of dynamics expressed by H2A.Z–H2B is more pronounced due to the participation of H2A.Z and supports our previous observation on the role of H2A.Z in bringing the nucleosome into a transcriptionally active state. The dynamics of the nucleosome dimer in the second slowest mode is comparatively constrained along the axis conjugate to the nucleosome dyad (Fig. 4.7I.B), which is already observed as the second slowest mode of the

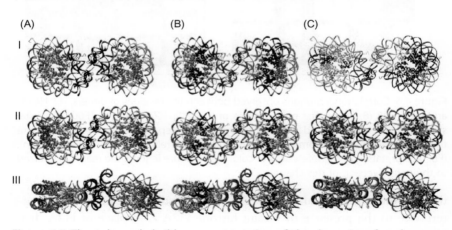

Figure 4.7 The color-coded ribbon representation of the dynamics of nucleosome dimers formed by 1EQZ (I) and 1F66 (II) in the first (A), second (B), and third (C) global modes. The regions graded from black to red represent the rigid to the flexible domains, respectively. The side view of the first three global modes of the dimer formed by 1F66 is shown in III. (For interpretation of the references to color in this figure legend, the reader is referred to the online version of this chapter.)

nucleosome monomer. Unlike the first global mode, the regions that acted as the hinge domain for the first global mode (the linker DNA/histone and the regions of H2A–H2B dimer interacting with the DNA sequences) are highly dynamic (Fig. 4.7II.B). The nucleosome with H2A.Z variant, the DNA–(H2A.Z–H2B) interacting regions (α2–L2–α3–αC regions of H2B and α2–L1–α1 of H2A.Z), expresses stronger dynamics than that of the histone H2A. In the third slowest mode, the fluctuations of both the canonical nucleosomes are almost frozen, whereas the nucleosomes with H2A.Z show equally dynamic conformation similar to the first two global modes, which reinforce again the role of H2A.Z in determining the transcriptionally active conformation of nucleosome. The pattern of flexible and rigid domains observed in this mode reflects the first global mode of dynamics observed with respect to the dyad axis in the nucleosome monomer. Altogether, the global dynamics of nucleosome dimer reveals the active participation of the entire domains (histone proteins, linker histone, and DNA) in the collective global motions as explained in the monomeric nucleosome.

Similar to the representation of the residual fluctuation in the nucleosome monomer, two conformations were generated by adding and subtracting the ANM-derived residue fluctuations (ANM conformation) and were superimposed with the crystal structure of the dimer. Superimposition of the crystal and ANM conformations revealed (i) conformational fluctuation perpendicular to the dyadic plane such that the H3–H4 dimer and linker DNA/histone is bent below the crystal conformation while the other half of the nucleosome is elevated above the crystal conformation and (ii) lateral contraction parallel to the dyad axis as expressed by the monomer (Fig. 4.8).

The second slowest mode of dynamics of the nucleosome dimer shows a coupled motion of both (i) fluctuations perpendicular to the dyadic plane as observed in the global mode and (ii) a planar twist with respect to the linker DNA–histone region (shown by the bend arrows in Fig. 4.9). The planar twist and lateral contraction/breathing motions of the DNA superhelix are likewise highly pronounced when compared to that of histone proteins. Roccatano, Barthel, and Zacharias (2007) have also reported such out-of-planar bending motion of nucleosome as one of the softest modes using molecular dynamics simulations. These functional dynamics observed in the first two slowest modes is highly pronounced in the nucleosome with H2A.Z histone. Such in-plane and out-of-plane motions develop distortions in histone–DNA interactions and weaken the stability of nucleosome and hence ultimately result in the nucleosome dissociation (Muthurajan

(A)   (B)

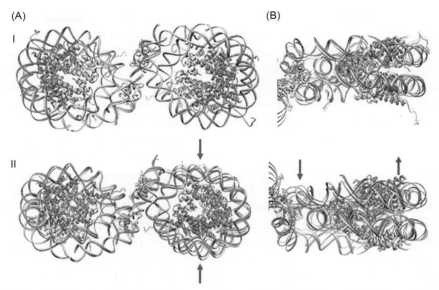

Figure 4.8 The conformational changes observed in the first slowest global mode of 1EQZ (I) and 1F66 (II) nucleosome dimer. Superposition of the modeled dimer (gray) with the conformation generated by adding the anisotropic fluctuation vector (magenta) is shown in (A) front and (B) side view with respect to the dyad of the right nucleosome. (For interpretation of the references to color in this figure legend, the reader is referred to the online version of this chapter.)

et al., 2003). The global dynamics of nucleosome complexes such as trimer and tetramer with only the variant histone H2A.Z is discussed further as the effective dynamics/influence of variant histone H2A.Z has already been proven in the dynamics of nucleosome monomer and dimer.

2.3. Nucleosome trimer

The 1F66 trimer expresses different mode of dynamics in first four slowest modes and each nucleosome significantly contributes for the functional dynamics (Fig. 4.10). In all these slowest modes, the nucleosome adopts different hinge domains for the respective global dynamics in a symmetric fashion. The global mode (first slowest mode) of 1F66 trimer (Fig. 4.10A) is highly governed by the dynamics of the first two nucleosomes, and the third nucleosome acts as the hinge nucleosome for the free motion of these two nucleosomes as its residue motions are highly conserved. In addition to this third nucleosome, the linker DNA and histones also act as the spatially conserved domains for the dynamics of the trimer. It is interesting to note that,

(A)   (B)

I

II

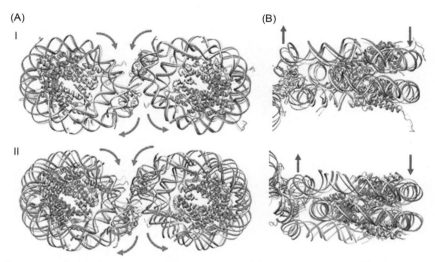

Figure 4.9 The conformational changes observed in the second slowest global mode of 1EQZ (I) and 1F66 (II) nucleosome dimer. Superposition of the modeled dimer (gray) with the conformation generated by adding the anisotropic fluctuation vector (magenta) is shown in (A) front and (B) side view with respect to the dyad of the right nucleosome. (For interpretation of the references to color in this figure legend, the reader is referred to the online version of this chapter.)

in both the active nucleosomes, the regions encompassed by the dimer of H2A.Z–H2B are highly dynamic (colored as red). In the second slowest mode (Fig. 4.10B), the domain formed by the linker DNA–proteins provides a highly fluctuating domain for the entire dynamics of the nucleosome trimer. All the three nucleosomes involve in a dynamic motions which are conjugate to the dyad axis as observed in the second slowest mode of the 1F66 dimer. The linker histone, that is, known to stabilize the association of histone octamer with the nucleosomal DNA, is highly active in this mode and may facilitate intranucleosomal transitions. In the third slowest mode (Fig. 4.10C), the middle nucleosome expresses rigidity along the dyad axis and the flexible regions flip with regard to this central hinge domain. The other two terminal nucleosomes express similar pattern of dynamics as observed in the second slowest mode of this trimer and facilitate nucleosome sliding due to their active linker histone–DNA domain. By keeping the middle nucleosome rigid, the terminal nucleosomes of the trimer express lateral motions with respect to the dyad axis without any nucleosomal displacement as the fourth slowest mode (Fig. 4.10D). It is interesting to mention that the trimer follows a collective pattern of dynamics expressed by the nucleosome dimer within its global mode.

(A)   (B)

(C)   (D)

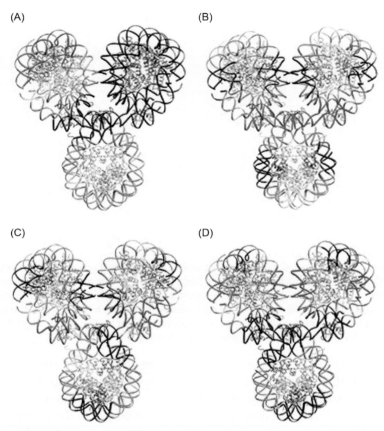

Figure 4.10 The color-coded ribbon representation of the dynamics of nucleosome tri-mer formed by 1F66 in the first (A), second (B), third (C), and (D) fourth global modes. The regions graded from black to red represent the rigid to the flexible domains, respectively. (For interpretation of the references to color in this figure legend, the reader is referred to the online version of this chapter.)

The conformational fluctuation of 1F66 trimer observed in the first two slowest modes is shown in Fig. 4.11 along with the superimposed crystal structure. Panel A shows the front view and B shows the side view in the first (I) and second (II) slowest modes. A combination of in-planar stretching (shown by arrows) and out-of-planar twisting (shown by curved arrows) motions is observed as the first two slow modes of the nucleosome tetramer. These fluctuations are similar in nature but differ in amplitude. These in-planar stretching and out-of-planar twisting motions are also evidenced as the relaxation mechanisms of the 1F66 dimer in the first two slow modes.

(A)   (B)

I

II

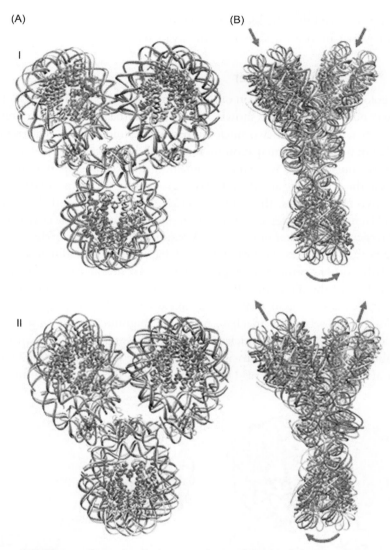

Figure 4.11 The conformational changes observed in the first two slowest global modes of 1F66 nucleosome trimer. Superposition of the modeled 1F66 trimer (gray) with the conformation generated by adding the anisotropic fluctuation vector (magenta) is shown in (A) front and (B) side view for the first (I) and second (II) slowest modes. (For interpretation of the references to color in this figure legend, the reader is referred to the online version of this chapter.)

It is also observed that the loci of the axis distinguishing both in-planar and out-of-planar motions lies ~45° to the dyad axis.

2.4. Nucleosome tetramer

The color-coded representation of the dynamics of 1F66 tetramer in the first and second slowest modes is shown in Fig. 4.12. In the global modes of tetramer, either the terminal two nucleosomes or the middle two nucleosomes involve in the active participation on functional dynamics while keeping the other two nucleosomes as the hinge domain, that is, keeping the two nucleosomes in the active state and the other two in a comparatively inactive state. In the active nucleosomes, the complete H2A.Z–H2B dimer and the part of H3–H4 tetramer that interacts with H2A.Z–H2B shows pronounced dynamics with respect to the linker histone–DNA region. The dynamics of these active nucleosomes replicates the first slowest global mode of the dimer and also reveals the role of symmetry in the global dynamics of the nucleosome tetramer.

Analysis of the directionality of the global dynamics of the nucleosome tetramer reveals (i) the in-planar stretching as well as out-of-planar bending motions as the relaxation mechanisms of the oligonucleosome and (ii) the freedom of the individual nucleosome in expressing the combination of the above-mentioned motions as the global mode of dynamics (Fig. 4.13). The active participation of DNA along with the interacting histone

(A)   (B)

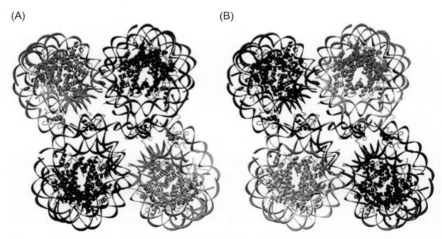

Figure 4.12 The color-coded ribbon diagram shows the dynamics of the 1F66 tetramer in the first (A) as well as second (B) global modes. The amplitude of fluctuation is graded as in Fig. 4.7. (For color version of this figure, the reader is referred to the online version of this chapter.)

(A)   (B)

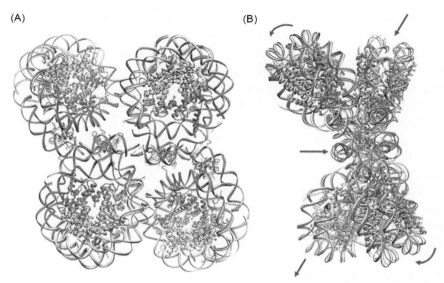

Figure 4.13 The conformational changes observed in the first global mode of 1F66 nucleosome tetramer. Superposition of the modeled 1F66 tetramer (gray) with the conformation generated by adding the anisotropic fluctuation vector (magenta) is shown in (A) front and (B) side view. (For interpretation of the references to color in this figure legend, the reader is referred to the online version of this chapter.)

octameric core reveals its ability in positioning the octameric core that result in nucleosome sliding/unwinding. The observed on and off states of histone H1 play a vital role in sustaining the nucleosomal complexes in the active/inactive state for transcription elongation and initiation (O'Neill, Meersseman, Pennings, & Bradbury, 1995). In general, the overall dynamics of the oligonucleosome is directly governed by the explicit motions of all domains.

From the structural dynamics of the modeled oligonucleosomes, it can be stated that the incorporation of H2A.Z histone releases the interresidual interactions and facilitates motions of higher amplitude for its functional dynamics in the global mode. Such vibrant dynamics emerged by the presence of H2A.Z would facilitate the oligonucleosome to be in its active conformation for further activation processes.

3. CONCLUSIONS AND OUTLOOK

Structural biology of chromatin fiber has become the topic of interest from the early 1980s since the understanding of eukaryotic genomes as complex of condensed, heterogeneous chromatin fibers. With the recent advent of computational approaches, understanding the biomolecular functions is extensively studied with less effort when compared to the experimental studies. Even though it is difficult to explain the role of H2A.Z in promoting the formation/destabilization of the 30-nm fiber from the analysis of the structural dynamics of the oligonucleosomes (up to tetramers), the release of interresidual interaction in the octameric histone core by the presence of H2A.Z and the emerging high amplitude motions (such as in-planar breathing motion along the dyad and out-of-planar bending motions) evidence the vibrant role of H2A.Z in gene regulation. In light of these findings, the role of variant histones from the structural-to-functional aspects in chromatin biology is explicitly revealed using computational approaches like coarse grained dynamics, and such conventional methods would facilitate the understanding of chromatin biology beyond the existing level of knowledge.

ACKNOWLEDGMENT

The authors would like to acknowledge Prof. Ivet Bahar, University of Pittsburgh for her valuable advices concerning the coarse grained dynamics.

REFERENCES

Abbott, D. W., Ivanova, V. S., Wang, X., Bonner, W. M., & Ausio, J. (2001). Characterization of the stability and folding of H2A.Z chromatin particles: Implications for transcriptional activation. *The Journal of Biological Chemistry, 276*, 41945–41949.

Adam, M., Robert, F., Larochelle, M., & Gaudreau, L. (2001). H2A.Z is required for global chromatin integrity and for recruitment of RNA polymerase II under specific conditions. *Molecular and Cellular Biology, 21*, 6270–6279.

Ahmad, K., & Henikoff, S. (2002a). The histone variant H3.3 marks active chromatin by replication-independent nucleosome assembly. *Molecular Cell, 9*, 1191–1200.

Ahmad, K., & Henikoff, S. (2002b). Histone H3 variants specify modes of chromatin assembly. *Proceedings of the National Academy of Sciences of the United States of America, 99*(Suppl. 4), 16477–16484.

Arents, G., Burlingame, R. W., Wang, B. C., Love, W. E., & Moudrianakis, E. N. (1991). The nucleosomal core histone octamer at 3.1 A resolution: A tripartite protein assembly and a left-handed superhelix. *Proceedings of the National Academy of Sciences of the United States of America, 88*, 10148–10152.

Arya, G., Maitra, A., & Grigoryev, S. A. (2010). A structural perspective on the where, how, why, and what of nucleosome positioning. *Journal of Biomolecular Structure & Dynamics, 27*, 803–820.

Atilgan, A. R., Durell, S. R., Jernigan, R. L., Demirel, M. C., Keskin, O., & Bahar, I. (2001). Anisotropy of fluctuation dynamics of proteins with an elastic network model. *Biophysical Journal, 80*, 505–515.

Axel, R. (1975). Cleavage of DNA in nuclei and chromatin with staphylococcal nuclease. *Biochemistry (Mosc), 14*, 2921–2925.

Bahar, I., Atilgan, A. R., & Erman, B. (1997). Direct evaluation of thermal fluctuations in proteins using a single-parameter harmonic potential. *Folding & Design, 2*, 173–181.

Bao, Y., Konesky, K., Park, Y. J., Rosu, S., Dyer, P. N., Rangasamy, D., et al. (2004). Nucleosomes containing the histone variant H2A.Bbd organize only 118 base pairs of DNA. *The EMBO Journal, 23*, 3314–3324.

Bonner, W. M., & Pollard, H. B. (1975). The presence of F3-a1 dimers and F1 oligomers in chromatin. *Biochemical and Biophysical Research Communications, 64*, 282–288.

Brown, D. T. (2001). Histone variants: Are they functionally heterogeneous? *Genome Biology, 2*(7), REVIEWS0006.1–REVIEWS0006.6.

Bryan, P. N., Wright, E. B., & Olins, D. E. (1979). Core nucleosomes by digestion of reconstructed histone-DNA complexes. *Nucleic Acids Research, 6*, 1449–1465.

Cairns, B. R. (2009). The logic of chromatin architecture and remodelling at promoters. *Nature, 461*, 193–198.

Celeste, A., Difilippantonio, S., Difilippantonio, M. J., Fernandez-Capetillo, O., Pilch, D. R., Sedelnikova, O. A., et al. (2003). H2AX haploinsufficiency modifies genomic stability and tumor susceptibility. *Cell, 114*, 371–383.

Cerf, C., Lippens, G., Ramakrishnan, V., Muyldermans, S., Segers, A., Wyns, L., et al. (1994). Homo- and heteronuclear two-dimensional NMR studies of the globular domain of histone H1: Full assignment, tertiary structure, and comparison with the globular domain of histone H5. *Biochemistry (Mosc), 33*, 11079–11086.

Chadwick, B. P., & Willard, H. F. (2001). A novel chromatin protein, distantly related to histone H2A, is largely excluded from the inactive X chromosome. *The Journal of Cell Biology, 152*, 375–384.

Chadwick, B. P., & Willard, H. F. (2002). Cell cycle-dependent localization of macroH2A in chromatin of the inactive X chromosome. *The Journal of Cell Biology, 157*, 1113–1123.

Changolkar, L. N., & Pehrson, J. R. (2002). Reconstitution of nucleosomes with histone macroH2A1.2. *Biochemistry (Mosc), 41*, 179–184.

Chung, S. Y., Hill, W. E., & Doty, P. (1978). Characterization of the histone core complex. *Proceedings of the National Academy of Sciences of the United States of America, 75*, 1680–1684.

Churikov, D., Siino, J., Svetlova, M., Zhang, K., Gineitis, A., Morton Bradbury, E., et al. (2004). Novel human testis-specific histone H2B encoded by the interrupted gene on the X chromosome. *Genomics, 84*, 745–756.

Clapier, C. R., & Cairns, B. R. (2009). The biology of chromatin remodeling complexes. *Annual Review of Biochemistry, 78*, 273–304.

Clarkson, M. J., Wells, J. R., Gibson, F., Saint, R., & Tremethick, D. J. (1999). Regions of variant histone His2AvD required for Drosophila development. *Nature, 399*, 694–697.

Costanzi, C., & Pehrson, J. R. (1998). Histone macroH2A1 is concentrated in the inactive X chromosome of female mammals. *Nature, 393*, 599–601.

Davie, J. R. (1998). Covalent modifications of histones: Expression from chromatin templates. *Current Opinion in Genetics & Development, 8*, 173–178.

Doenecke, D., Albig, W., Bouterfa, H., & Drabent, B. (1994). Organization and expression of H1 histone and H1 replacement histone genes. *Journal of Cellular Biochemistry, 54*, 423–431.

Dorigo, B., Schalch, T., Bystricky, K., & Richmond, T. J. (2003). Chromatin fiber folding: Requirement for the histone H4 N-terminal tail. *Journal of Molecular Biology, 327*, 85–96.

Draves, P. H., Lowary, P. T., & Widom, J. (1992). Co-operative binding of the globular domain of histone H5 to DNA. *Journal of Molecular Biology, 225*, 1105–1121.

Eickbush, T. H., & Moudrianakis, E. N. (1978). The histone core complex: An octamer assembled by two sets of protein-protein interactions. *Biochemistry (Mosc), 17*, 4955–4964.

Eickbush, T. H., Watson, D. K., & Moudrianakis, E. N. (1976). A chromatin-bound proteolytic activity with unique specificity for histone H2A. *Cell, 9*, 785–792.

Fan, J. Y., Gordon, F., Luger, K., Hansen, J. C., & Tremethick, D. J. (2002). The essential histone variant H2A.Z regulates the equilibrium between different chromatin conformational states. *Nature Structural Biology, 9*, 172–176.

Finch, J. T., & Klug, A. (1976). Solenoidal model for superstructure in chromatin. *Proceedings of the National Academy of Sciences of the United States of America, 73*, 1897–1901.

Garcia, B. A., Barber, C. M., Hake, S. B., Ptak, C., Turner, F. B., Busby, S. A., et al. (2005). Modifications of human histone H3 variants during mitosis. *Biochemistry (Mosc), 44*, 13202–13213.

Greil, W., Igo-Kemenes, T., & Zachau, H. G. (1976). Nuclease digestion in between and within nucleosomes. *Nucleic Acids Research, 3*, 2633–2644.

Grigoryev, S. A., Arya, G., Correll, S., Woodcock, C. L., & Schlick, T. (2009). Evidence for heteromorphic chromatin fibers from analysis of nucleosome interactions. *Proceedings of the National Academy of Sciences of the United States of America, 106*, 13317–13322.

Haliloglu, T., Bahar, I., & Erman, B. (1997). Gaussian dynamics of folded proteins. *Physical Review Letters, 79*, 3090–3093.

Harp, J. M., Hanson, B. L., Timm, D. E., & Bunick, G. J. (2000). Asymmetries in the nucleosome core particle at 2.5 angstrom resolution. *Acta Crystallographica. Section D, Biological Crystallography, 56*, 1513–1534.

Hayes, J. J., Clark, D. J., & Wolffe, A. P. (1991). Histone contributions to the structure of DNA in the nucleosome. *Proceedings of the National Academy of Sciences of the United States of America, 88*, 6829–6833.

Hayes, J. J., Tullius, T. D., & Wolffe, A. P. (1990). The structure of DNA in a nucleosome. *Proceedings of the National Academy of Sciences of the United States of America, 87*, 7405–7409.

Hendzel, M. J., & Davie, J. R. (1990). Nucleosomal histones of transcriptionally active/competent chromatin preferentially exchange with newly synthesized histones in quiescent chicken erythrocytes. *The Biochemical Journal, 271*, 67–73.

Horn, P. J., & Peterson, C. L. (2002). Molecular biology. Chromatin higher order folding—Wrapping up transcription. *Science, 297*, 1824–1827.

Horowitz, R. A., Koster, A. J., Walz, J., & Woodcock, C. L. (1997). Automated electron microscope tomography of frozen-hydrated chromatin: The irregular three-dimensional zigzag architecture persists in compact, isolated fibers. *Journal of Structural Biology, 120*, 353–362.

Isenberg, I. (1979). Histones. *Annual Review of Biochemistry, 48*, 159–191.

Jackson, J. D., Falciano, V. T., & Gorovsky, M. A. (1996). A likely histone H2A.F/Z variant in Saccharomyces cerevisiae. *Trends in Biochemical Sciences, 21*, 466–467.

Jackson, J. D., & Gorovsky, M. A. (2000). Histone H2A.Z has a conserved function that is distinct from that of the major H2A sequence variants. *Nucleic Acids Research, 28*, 3811–3816.

Johnson, L. M., Fisher-Adams, G., & Grunstein, M. (1992). Identification of a non-basic domain in the histone H4 N-terminus required for repression of the yeast silent mating loci. *The EMBO Journal, 11*, 2201–2209.

Kornberg, R. D. (1977). Structure of chromatin. *Annual Review of Biochemistry, 46*, 931–954.

Kouzarides, T. (2007). Chromatin modifications and their function. *Cell, 128*, 693–705.

Leach, A. R. (2001). *Molecular modelling: Principles and applications.* Upper Saddle River: Prentice Hall.

Lowell, J. E., Kaiser, F., Janzen, C. J., & Cross, G. A. (2005). Histone H2AZ dimerizes with a novel variant H2B and is enriched at repetitive DNA in Trypanosoma brucei. *Journal of Cell Science, 118*, 5721–5730.

Luger, K., Mader, A. W., Richmond, R. K., Sargent, D. F., & Richmond, T. J. (1997). Crystal structure of the nucleosome core particle at 2.8 A resolution. *Nature, 389*, 251–260.

Luger, K., Rechsteiner, T. J., & Richmond, T. J. (1999a). Preparation of nucleosome core particle from recombinant histones. *Methods in Enzymology, 304*, 3–19.

Luger, K., Rechsteiner, T. J., & Richmond, T. J. (1999b). Expression and purification of recombinant histones and nucleosome reconstitution. *Methods in Molecular Biology, 119*, 1–16.

Lutter, L. C. (1979). Precise location of DNase I cutting sites in the nucleosome core determined by high resolution gel electrophoresis. *Nucleic Acids Research*, 6, 41–56.

Malik, H. S., & Henikoff, S. (2003). Phylogenomics of the nucleosome. *Nature Structural Biology*, 10, 882–891.

McGhee, J. D., & Felsenfeld, G. (1980). Nucleosome structure. *Annual Review of Biochemistry*, 49, 1115–1156.

McGhee, J. D., Nickol, J. M., Felsenfeld, G., & Rau, D. C. (1983). Higher order structure of chromatin: Orientation of nucleosomes within the 30 nm chromatin solenoid is independent of species and spacer length. *Cell*, 33, 831–841.

Mohd-Sarip, A., & Verrijzer, C. P. (2004). Molecular biology. A higher order of silence. *Science*, 306, 1484–1485.

Muller, M., Katsov, K., & Schick, M. (2006). Biological and synthetic membranes: What can be learned from a coarse-grained description? *Physics Reports: Review Section of Physics Letters*, 434, 113–176.

Muthurajan, U. M., Park, Y. J., Edayathumangalam, R. S., Suto, R. K., Chakravarthy, S., Dyer, P. N., et al. (2003). Structure and dynamics of nucleosomal DNA. *Biopolymers*, 68, 547–556.

Nielsen, S. O., Lopez, C. F., Srinivas, G., & Klein, M. L. (2004). Coarse grain models and the computer simulation of soft materials. *Journal of Physics. Condensed Matter*, 16, R481–R512.

Noll, M., & Kornberg, R. D. (1977). Action of micrococcal nuclease on chromatin and the location of histone H1. *Journal of Molecular Biology*, 109, 393–404.

O'Neill, T. E., Meersseman, G., Pennings, S., & Bradbury, E. M. (1995). Deposition of histone H1 onto reconstituted nucleosome arrays inhibits both initiation and elongation of transcripts by T7 RNA polymerase. *Nucleic Acids Research*, 23, 1075–1082.

Palmer, D. K., O'Day, K., Trong, H. L., Charbonneau, H., & Margolis, R. L. (1991). Purification of the centromere-specific protein CENP-A and demonstration that it is a distinctive histone. *Proceedings of the National Academy of Sciences of the United States of America*, 88, 3734–3738.

Palmer, D. K., O'Day, K., Wener, M. H., Andrews, B. S., & Margolis, R. L. (1987). A 17-kD centromere protein (CENP-A) copurifies with nucleosome core particles and with histones. *The Journal of Cell Biology*, 104, 805–815.

Pehrson, J. R., & Fuji, R. N. (1998). Evolutionary conservation of histone macroH2A subtypes and domains. *Nucleic Acids Research*, 26, 2837–2842.

Prunell, A., Kornberg, R. D., Lutter, L., Klug, A., Levitt, M., & Crick, F. H. (1979). Periodicity of deoxyribonuclease I digestion of chromatin. *Science*, 204, 855–858.

Ramakrishnan, V. (1997). Histone structure and the organization of the nucleosome. *Annual Review of Biophysics and Biomolecular Structure*, 26, 83–112.

Ramaswamy, A., Bahar, I., & Ioshikhes, I. (2005). Structural dynamics of nucleosome core particle: Comparison with nucleosomes containing histone variants. *Proteins*, 58, 683–696.

Ramaswamy, A., & Ioshikhes, I. (2007). Global dynamics of newly constructed oligonucleosomes of conventional and variant H2A.Z histone. *BMC Structural Biology*, 7, 76.

Richmond, T. J., Finch, J. T., Rushton, B., Rhodes, D., & Klug, A. (1984). Structure of the nucleosome core particle at 7 A resolution. *Nature*, 311, 532–537.

Richmond, T. J., & Widom, J. (2000). Nucleosome and chromatin structure. In S. C. R. Elgin & J. L. Workman (Eds.), *Chromatin structure and gene expression*. Oxford, UK: Oxford University Press.

Roccatano, D., Barthel, A., & Zacharias, M. (2007). Structural flexibility of the nucleosome core particle at atomic resolution studied by molecular dynamics simulation. *Biopolymers*, 85, 407–421.

Santisteban, M. S., Kalashnikova, T., & Smith, M. M. (2000). Histone H2A.Z regulates transcription and is partially redundant with nucleosome remodeling complexes. *Cell, 103,* 411–422.

Schalch, T., Duda, S., Sargent, D. F., & Richmond, T. J. (2005). X-ray structure of a tetranucleosome and its implications for the chromatin fibre. *Nature, 436,* 138–141.

Schlick, T., Collepardo-Guevara, R., Halvorsen, L. A., Jung, S., & Xiao, X. (2011). Biomolecular modeling and simulation: A field coming of age. *Quarterly Reviews of Biophysics, 44,* 191–228.

Shaw, B. R., Herman, T. M., Kovacic, R. T., Beaudreau, G. S., & Van Holde, K. E. (1976). Analysis of subunit organization in chicken erythrocyte chromatin. *Proceedings of the National Academy of Sciences of the United States of America, 73,* 505–509.

Shillcock, J. C., & Lipowsky, R. (2006). The computational route from bilayer membranes to vesicle fusion. *Journal of Physics. Condensed Matter, 18,* S1191–S1219.

Shires, A., Carpenter, M. P., & Chalkley, R. (1976). A cysteine-containing H2B-like histone found in mature mammalian testis. *The Journal of Biological Chemistry, 251,* 4155–4158.

Sierra, F., Stein, G., & Stein, J. (1983). Structure and in vitro transcription of a human H4 histone gene. *Nucleic Acids Research, 11,* 7069–7086.

Simpson, R. T. (1978). Structure of the chromatosome, a chromatin particle containing 160 base pairs of DNA and all the histones. *Biochemistry (Mosc), 17,* 5524–5531.

Simpson, R. T., & Kunzler, P. (1979). Cromatin and core particles formed from the inner histones and synthetic polydeoxyribonucleotides of defined sequence. *Nucleic Acids Research, 6,* 1387–1415.

Simpson, R. T., & Whitlock, J. P., Jr. (1976). Chemical evidence that chromatin DNA exists as 160 base pair beads interspersed with 40 base pair bridges. *Nucleic Acids Research, 3,* 117–127.

Sivolob, A., Lavelle, C., & Prunell, A. (2003). Sequence-dependent nucleosome structural and dynamic polymorphism. Potential involvement of histone H2B N-terminal tail proximal domain. *Journal of Molecular Biology, 326,* 49–63.

Sollner-Webb, B., & Felsenfeld, G. (1975). A comparison of the digestion of nuclei and chromatin by staphylococcal nuclease. *Biochemistry (Mosc), 14,* 2915–2920.

Suto, R. K., Clarkson, M. J., Tremethick, D. J., & Luger, K. (2000). Crystal structure of a nucleosome core particle containing the variant histone H2A.Z. *Nature Structural Biology, 7,* 1121–1124.

Tachiwana, H., Osakabe, A., Kimura, H., & Kurumizaka, H. (2008). Nucleosome formation with the testis-specific histone H3 variant, H3t, by human nucleosome assembly proteins in vitro. *Nucleic Acids Research, 36,* 2208–2218.

Tagami, H., Ray-Gallet, D., Almouzni, G., & Nakatani, Y. (2004). Histone H3.1 and H3.3 complexes mediate nucleosome assembly pathways dependent or independent of DNA synthesis. *Cell, 116,* 51–61.

Thoma, F., Koller, T., & Klug, A. (1979). Involvement of histone H1 in the organization of the nucleosome and of the salt-dependent superstructures of chromatin. *The Journal of Cell Biology, 83,* 403–427.

Thomas, J. O. (1999). Histone H1: Location and role. *Current Opinion in Cell Biology, 11,* 312–317.

Thomas, J. O., Rees, C., & Finch, J. T. (1992). Cooperative binding of the globular domains of histones H1 and H5 to DNA. *Nucleic Acids Research, 20,* 187–194.

Van Holde, K. E. (1989). *Chromatin*. New York: Springer-Verlag.

Van Holde, K. E., Sahasrabuddhe, C. G., & Shaw, B. R. (1974). A model for particulate structure in chromatin. *Nucleic Acids Research, 1,* 1579–1586.

Varshavsky, A. J., Bakayev, V. V., & Georgiev, G. P. (1976). Heterogeneity of chromatin subunits in vitro and location of histone H1. *Nucleic Acids Research, 3,* 477–492.

Venturoli, M., Sperotto, M. M., Kranenburg, M., & Smit, B. (2006). Mesoscopic models of biological membranes. *Physics Reports: Review Section of Physics Letters*, *437*, 1–54.

Widom, J., & Klug, A. (1985). Structure of the 300A chromatin filament: X-ray diffraction from oriented samples. *Cell*, *43*, 207–213.

Wiedemann, S. M., Mildner, S. N., Bonisch, C., Israel, L., Maiser, A., Matheisl, S., et al. (2010). Identification and characterization of two novel primate-specific histone H3 variants, H3.X and H3.Y. *The Journal of Cell Biology*, *190*, 777–791.

Williams, S. P., Athey, B. D., Muglia, L. J., Schappe, R. S., Gough, A. H., & Langmore, J. P. (1986). Chromatin fibers are left-handed double helices with diameter and mass per unit length that depend on linker length. *Biophysical Journal*, *49*, 233–248.

Wolffe, A. (1998). *Chromatin: Structure and function*. San Diego: Academic Press.

Woodcock, C. L., Frado, L. L., & Rattner, J. B. (1984). The higher-order structure of chromatin: Evidence for a helical ribbon arrangement. *The Journal of Cell Biology*, *99*, 42–52.

Worcel, A., Strogatz, S., & Riley, D. (1981). Structure of chromatin and the linking number of DNA. *Proceedings of the National Academy of Sciences of the United States of America*, *78*, 1461–1465.

Workman, J. L., & Kingston, R. E. (1998). Alteration of nucleosome structure as a mechanism of transcriptional regulation. *Annual Review of Biochemistry*, *67*, 545–579.

Yuan, J., Adamski, R., & Chen, J. (2010). Focus on histone variant H2AX: To be or not to be. *FEBS Letters*, *584*, 3717–3724.

Zheng, C., & Hayes, J. J. (2003). Structures and interactions of the core histone tail domains. *Biopolymers*, *68*, 539–546.

Zhou, Y. B., Gerchman, S. E., Ramakrishnan, V., Travers, A., & Muyldermans, S. (1998). Position and orientation of the globular domain of linker histone H5 on the nucleosome. *Nature*, *395*, 402–405.

Zlatanova, J., Bishop, T. C., Victor, J. M., Jackson, V., & van Holde, K. (2009). The nucleosome family: Dynamic and growing. *Structure*, *17*, 160–171.

Zweidler, A. (1984). Core histone variants of the mouse: Primary structure and differential expression. In G. S. Stein, J. L. Stein & W. F. Marzluff (Eds.), *Histone genes: Structure, organization, and regulation* (pp. 339–371). New York: John Wiley & Sons.

CHAPTER FIVE

Unconventional Actin Configurations Step into the Limelight

Unai Silván[*,1], Brigitte M. Jockusch[†], Cora-Ann Schoenenberger[*,2]

[*]Focal Area Structural Biology and Biophysics, Biozentrum, University of Basel, Basel, Switzerland
[†]Cell Biology, Zoological Institute, Technical University of Braunschweig, Braunschweig, Germany
[1]Present address: Uniklinik Balgrist, Orthopedic Biomechanics ETH Zurich, Institute for Biomechanics, Forchstrasse 340, CH-8008 Zurich, Switzerland
[2]Corresponding author: e-mail address: cora-ann.schoenenberger@unibas.ch

Contents

Abstract

The existence of a cellular machinery that is based on the reversible polymerization of globular nucleotide-bound protomers into polar microfilaments is a persistent feature from prokaryotes to higher vertebrates. However, while in bacteria, actin-like proteins with such properties have evolved into a large family with divergent sequences and polymeric structures, eukaryotes express only a small number of highly conserved actins. Indeed, the sequence of actin is one of the best conserved among eukaryotes and yet actin carries out many different functions at distinct cellular sites. Because of the notorious conservation and lack of suitable tools to examine structural plasticity, the vast majority of studies on cellular actin functions consider mainly two structural states, G-actin and F-actin. However, there is more to the structural plasticity of actin than first meets the eye. On one hand, more than 200 actin-binding proteins shape the conformation of actin and thereby regulate functional diversity. On the other hand, unconventional actin conformations that differ from monomeric G-actin are stepping into the limelight. In addition, supramolecular actin structures that extend beyond classical F-actin are emerging. Herein, we recapitulate the current knowledge on the structure and conformations of monomeric actin and its polymerization into higher order structures, paying special attention to less known forms and their involvement in actin function.

1. INTRODUCTION

Globular proteins that polymerize into filamentous structures to generate physical force are present from bacteria to higher eukaryotes. One well-studied example is the actin-like proteins (Alps) that belong to the actin-like ATPase domain superfamily, the members of which have a common core fold or "actin fold" and catalyze phosphoryl transfer or hydrolysis of ATP (Bork, Sander, & Valencia, 1992; Holmes, Sander, & Valencia, 1993; Kabsch & Holmes, 1995). Actin itself was first purified as an actomyosin complex in the early 1940s by Straub who already noted that this protein can exist in two states: filamentous (F-actin) and monomeric or globular (G-actin) (Straub, 1942). Together with Feuer, he also demonstrated the ionic strength dependency of the globular to filamentous transition *in vitro* (Straub & Feuer, 1950). The prokaryotic ancestors of actin were identified much later (Jones, Carballido-López, & Errington, 2001; Van den Ent, Amos, & Löwe, 2001), after solving the crystal structure of muscle actin (Kabsch, Mannherz, Suck, Pai, & Holmes, 1990) made it clear that actin belonged to a large family of actin-related proteins which share a structural feature, the "actin fold" (Bork et al., 1992). Only when structural features rather than sequence similarities were considered, it was appreciated that purified MreB from *Thermotoga maritima* has a structure remarkably similar to that of actin and can undergo actin-like polymerization.

However, while in bacteria, Alps rapidly evolved into a large family with diverse functions in eukaryotic cells, actin remained remarkably unchanged. From yeast to higher plants and mammals, actin has one of the most highly conserved amino acid sequences, with 90% identity between yeast and man. The number of actin genes increased substantially during evolution (there is 1 actin gene in yeast (Gallwitz & Seidel, 1980), 17 in *Dictyostelium* (Kindle &

Firtel, 1978), and 1 in man (Miwa et al., 1991)), but all their products differ only slightly in sequence and thus in structure. It has been convincingly argued that a conservative evolutionary pressure is exerted by the numerous actin–interacting proteins which control polymerization (Popp & Robinson, 2011). These interactions, together with its inherent plasticity, render actin one of the most versatile proteins in the eukaryotic cell, both with respect to structure and function.

2. STRUCTURAL ORGANIZATION OF ACTIN

2.1. G-Actin

Due to its propensity to form noncrystalline filamentous aggregates or par-acrystalline arrays under the conditions normally used for crystallization (Bremer & Aebi, 1992), monomeric actin has for long resisted crystallization attempts. Different strategies employed to prevent actin from polymerizing, including point mutations, covalent modification, or association with other proteins or ligands, finally allowed its structure to be determined at atomic resolution (Fig. 5.1; Kabsch et al., 1990). The first crystal structure of an actin subunit that had not been chemically modified or complexed with an actin-binding protein was reported in 2006 (Klenchin, Khaitlina, & Rayment,

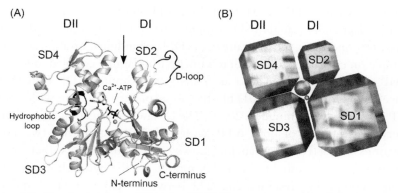

Figure 5.1 Atomic structure of G-actin. (A) Ribbon representation of the actin molecule with bound ATP and Ca^{2+} in which the cleft between the two major domains DI and DII is indicated by an arrow (Kabsch et al., 1990; drawn from PDB coordinates 1ATN). DNase-I binding loop (D-loop; residues 40–50) and hydrophobic loop (residues 262–272) are highlighted. (B) Schematic representation of subdomains 1–4 (SD1–SD4) with sizes according their relative mass. (For color version of this figure, the reader is referred to the online version of this chapter.)

2006). Today, more than 80 different crystal structures are known which slightly vary based on modification strategies or crystallization conditions.

The crystallographic studies revealed actin as a globular protein with a nearly flat shape and approximate dimensions of $6 \times 4 \times 2$ nm. As illustrated in Fig. 5.1, the molecule is conventionally described to comprise two major domains, DI and DII, which in turn are divided into two subdomains each (SD1–SD4). There is a nucleotide binding site in the deep cleft between DI and DII, with the nucleotide interacting with all four subdomains and a high affinity binding site for a cation. Considerable effort has been devoted to investigating how the bound nucleotide influences the conformation and activity of the protein. Variations in the width of the nucleotide binding cleft also depend on crystallization conditions. In most crystal structures of actin, the nucleotide binding cleft is closed around either ATP or ADP. In certain crystals, the cleft is slightly wider ("open"), for example, in crystals of ATP-β-actin bound to profilin, where the so-called P1 and P2 loops are separated by an additional 3 Å from closed conformations (Chik, Lindberg, & Schutt, 1996). However, molecular dynamic simulations showed that the open nucleotide pocket of the profilin/actin X-ray structure is unstable and closes in the absence of profilin (Minehardt, Kollman, Cooke, & Pate, 2006). Different states have been used to model actin filaments from electron microscopy reconstructions in the ADP and ADP-Pi conformations, respectively (Belmont, Orlova, Drubin, & Egelman, 1999). A highly flexible and mobile loop, named DNase-I binding loop (or D-loop) due to its high affinity for this enzyme, is located in SD2 (residues 40–50). Depending on the crystallization conditions, the structure of the DNase binding loop also varies. For example, in crystals of actin with tetramethylrhodamine coupled to Cys374 (TMR-actin), it is folded into an α-helix with bound ADP (Otterbein, Graceffa, & Dominguez, 2001) but has no secondary structure with bound ATP (Graceffa & Dominguez, 2003). The DNase binding loop is disordered in crystal structures of both ATP–actin and ADP–actin with two point mutations in SD4 (Rould, Wan, Joel, Lowey, & Trybus, 2006). Other regions that have been suggested to be highly mobile are the hydrophobic loop or "plug," comprising residues 262–272 between SD3 and SD4, and the N-terminal region located in SD1.

Under physiological salt conditions, actin polymerizes into filaments. This transition is an integral part of the function of actin in eukaryotic cells, and the equilibrium between monomeric G-actin and filamentous F-actin species depends on the cellular needs. Hence, *in vivo* actin is bound to ABPs that regulate its polymerization by locking the molecule in specific conformations. Although pivotal to cellular actin functions, the conformational changes associated with the transition between monomeric and polymeric forms are not fully understood, as so far there is no high-resolution structure for filamentous actin.

2.2. F-Actin

Eukaryotic F-actin is constructed from two protofilaments that gently wind around each other to form a polar helical polymer. Several bacterial Alps are also known to form F-actin-like helical arrangements from two protofilaments, yet with varied helical geometries (reviewed in Popp & Robinson, 2011). Recently, a unique filament architecture constructed from four protofilaments was reported for the protein Alp12 from *Clostridium tetani* (Popp et al., 2012). Based on these variations, one can anticipate more unknown actin structures to be identified in prokaryotes in the future.

For many years, eukaryotic F-actin was considered to be in a single conformational state with a highly ordered, homogeneous structure. However, recent studies reveal a rather high degree of heterogeneity along different filament segments, making it clear that F-actin cannot be described by a single structural model (Fan, Saunders, & Voth, 2012; Galkin, Orlova, Schröder, & Egelman, 2010). Hence, obtaining high-resolution atomic data on the different structural states of F-actin, in particular on the relative orientation of the actin protomers, remains extremely challenging.

The first atomic filament model was developed by fitting the atomic structure of the actin monomer to X-ray fiber diagrams of oriented F-actin gels (Holmes, Popp, Gebhard, & Kabsch, 1990) and was further refined in 1993 (Lorenz, Popp, & Holmes, 1993). In the Holmes or "G-like" model (Saunders & Voth, 2012), actin SD1 and SD2 are found in the outer part of the filament, while SD3 and SD4 are on the inside nearer to the filament axis. In more recent models (Fujii, Iwane, Yanagida, & Namba, 2010; Oda, Iwasa, Aihara, Maéda, & Narita, 2009), the actin subunit is flattened relative to G-actin, which has been suggested to facilitate intersubunit interactions. Furthermore, experimental evidence has shown that actin subunits can adopt multiple configurations (Galkin et al., 2010; Oztug Durer, Diraviyam, Sept, Kudryashov, & Reisler, 2010). Using frozen-hydrated filaments, Galkin et al. (2010) described six distinct "modes" that represent different filament structures. While the position of SD3 and SD4 was similar in all six modes, the configuration of SD2 was found to be highly variable (Galkin et al., 2010). In fact, SD2 contains the highly polymorphic DNase-I binding loop that can adopt either beta-strand or alpha helix conformations, as revealed by X-ray crystallography (Kabsch et al., 1990; Otterbein et al., 2001). It has been suggested that these structural changes might control connectivity between the two helical strands. A recent study using coarse-grained analysis of molecular dynamics simulations of the G-like (Holmes et al., 1990), the Oda (Oda et al., 2009), and the Namba (Fujii et al., 2010) model determined that a key motion in the G- to F-actin transition is the rotation of SD2 (Saunders & Voth, 2012).

Further information about the arrangement of monomers within the filament has been obtained from the crystallization of cross-linked actin oligomers (Dawson, Sablin, Spudich, & Fletterick, 2003). Recurrence of a specific interface in several species that are captured by different cross-linkers and independently crystallized has been taken as an indication of its presence in the filament. For example, based on a comparison of longitudinally cross-linked dimers, Sawaya et al. (2008) determined regions that presumably interact with each other in the F-actin state. However, in these models, the subunits retained the same conformation as in G-actin crystals, and hence, the structural plasticity of actin monomers in the F-state was not taken into account.

Recently, using filaments oriented in a gel, Oda et al. (2009) proposed a filament model at much higher detail than the previous ones. The Oda model agrees with earlier models with respect to the overall structure of the filament. In addition, it allows for the description of significant differences in the subdomain arrangement within the actin subunits in G- versus F-state (Oda et al., 2009). The major conformational change consists in a relative rotation of actin DI and DII by about 20° to each other. This results in a flattening of the monomer, which Oda and colleagues proposed to be related to the ATPase activity of actin. In addition, the movement of the molecule generates changes in the intramolecular interactions between the two major domains and facilitates interstrand connections to other actin subunits. For example, the hydrophobic loop, which is found between SD3 and SD4, changes its position and inserts into a pocket of the opposite strand, thereby stabilizing the filament structure (Scoville et al., 2006). The removal of the nucleotide from G-actin seems to induce the formation of a "flat conformation," which might explain the enhanced polymerization ability of nucleotide-free actin (De La Cruz et al., 2000). Although the significance of such domain motions within the filament is not yet resolved, a certain flexibility of the domains seems to be crucial, since binding of the toxin latrunculin to the cleft between the two domains blocks the structure in a flat conformation but strongly inhibits polymerization (Morton, Ayscough, & McLaughlin, 2000; Oda et al., 2009).

Similar to the G–like model for the F–actin structure (Holmes et al., 1990), the Oda model defines actin filaments either as a one-strand left-handed genetic helix of 5.9-nm pitch or as two-start, right-handed helices with a pitch of 72 nm. These measurements are approximate, since the helical twist of the filament naturally shows a high variation (Aebi, Millonig, Salvo, & Engel, 1986; Hanson, 1967). The fungal toxin phalloidin, together with specific ABPs (e.g., myosin), have been reported to reduce the variability of the twist angle, albeit within the "normal range." In contrast, members of the cofilin/ADF family of ABPs have been shown to dramatically reduce the twist of actin filaments (McGough, Pope, Chiu, & Weeds, 1997). This structural modification seems to be directly related to the severing activity of these ABPs, as it causes a weakening of the interstrand contacts of the filament (McGough & Chiu, 1999). Moreover, cofilin interactions prevent the binding of phalloidin, as the reduced twist angle is likely to result in the disruption of its binding site (Oda, Namba, & Maéda, 2005; Steinmetz et al., 1998).

3. TOOLS TO STUDY ACTIN LOCALIZATION, ARRANGEMENT, AND CONFORMATION

In the early years of actin research, the filament was considered the functional form of actin. In fact, localization studies predominantly focused on F–actin, not least because of the tools available. By far, the most widely used probe to detect filamentous actin is the mushroom toxin phalloidin, as first published by Wulf, Deboben, Bautz, Faulstich, and Wieland (1979). Fluorescent analogs of the bicyclic heptapeptide have been useful in localizing actin filaments in living or fixed cells as well as for visualizing individual actin filaments *in vitro*. As illustrated in Fig. 5.2, fluorescent phalloidin shows discrete staining patterns in different types of cells. However, phalloidin has been shown to interfere with the distribution of filamentous actin (Cooper, 1987). More importantly, phalloidin binding involves a specific F–actin conformation of several consecutive actin subunits (Oda et al., 2009; Steinmetz et al., 1998), and thus, actin configurations consisting of only short or not F–like assemblies may evade detection by this toxin in the cell.

3.1. Antibodies

The most versatile tools to study the localization of distinct actin species in cells are antibodies. Because the sequence of actin is notoriously conserved, actin is a poor antigen and the generation of actin antibodies has been a challenge. Lazarides and Weber (1974) reported the first specific antibody that

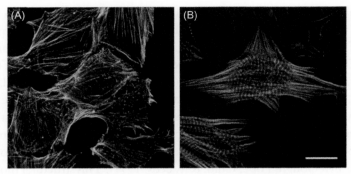

Figure 5.2 Fluorescent phalloidin binds to F-actin with high affinity and unambiguously reveals this form of actin in cells. (A) In 2D cultures of Rat-2 fibroblasts, the most evident phalloidin-positive structures are stress fibers. (B) Phalloidin binding to thin filaments in neonatal rat cardiomyocytes reveals the periodic pattern of the sarcomere organization. Scale bar represents 20 μm. (For color version of this figure, the reader is referred to the online version of this chapter.)

was reliably detecting cytoplasmic actin in mammalian cells in 1974. A small number of antibodies that bind specific isoforms have been established by immunizing with peptides representing the N-terminal actin isoform sequence (e.g., Gimona et al., 1994; Skalli et al., 1986). Different approaches have been used to overcome its low antigenicity, including immunization with denatured actin (Lazarides & Weber, 1974) or with complexes of actin with ABPs (Gonsior et al., 1999), exploiting the repetitive display of actin sequences engineered onto the surface of self-assembling peptidic nanoparticles (Schroeder et al., 2009), and chemically cross-linking unconventional arrangements of actin (Schoenenberger et al., 2005; Silván et al., 2012). These approaches have remarkably increased the number of probes available for the detection and study of actin *in vivo* and *in vitro*. Nowadays, it is possible to discern not only between G- and F-actin but as well between different conformations of the actin subunits.

One approach to generate antibodies that detect the nonpolymerized form of actin in cells was to use the hydrophobic loop as immunogen (Schroeder et al., 2009). Since this structure is buried in the filament, it is not readily accessible to loop antibodies and they primarily bind to the nonfilamentous form of actin. Notably, denaturing of filaments by treating cells with ice-cold methanol exposes this loop and allows the staining of stress fibers, that is, bundles of filaments that are the most prominent supramolecular actin structures in cultured vertebrate cells (Fig. 5.3A). An alternative approach was to immunize mice with an actin–profilin complex, which is

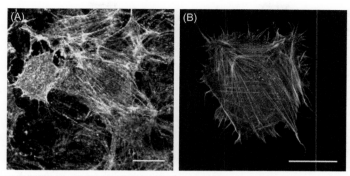

Figure 5.3 Postfixation and *in situ* fluorescent labeling of the actin cytoskeleton in cells. (A) Immunolabeling of methanol-treated fibroblasts with the monoclonal antibody 1F1 (Schroeder et al., 2009) results in a stress fiber pattern similar to that observed with fluorescent phalloidin. However, the antibody also detects actin that is not incorporated in stress fibers. (B) Transfection of cells with Lifeact fused to mGFP allows monitoring monomeric and F-actin in living cells (Riedl et al., 2008). Note the abundance of filopodial extensions in HeLa cells that are transiently transfected with mGFP-Lifeact. Scale bars represent 20 μm. (For color version of this figure, the reader is referred to the online version of this chapter.)

an abundant form of monomeric actin in the cell (Gonsior et al., 1999). Moreover, through its interaction with profilin, actin assumes a conformation that is unlikely to be present in native F–actin. One of the resulting monoclonal antibodies called 2G2 stained a specific nonfilamentous form of actin in a number of cell types. Epitope mapping of 2G2 revealed its reactivity with a conformational, three–partite epitope at the bottom of the cleft in the G-actin structure. One could imagine that the binding of profilin opens the G–actin fold (Chik et al., 1996) and thereby exposes the bottom of the cleft to the immune system of the injected host. Immunolabeling revealed the ability of this monoclonal antibody to detect a specific population of actin present in the nucleus of several mammalian cells (see Fig. 5.4B; Gonsior et al., 1999; Schoenenberger et al., 2005). In addition, specific fixation and permeabilization procedures seem to unmask the respective epitopes, and consequently, 2G2-reactive actin structures were identified at other cellular sites. For example, 2G2-positive actin has been found at the cell periphery (Fig. 5.7C) but also in filopodial tips and stress fibers (Schoenenberger et al., 2005).

A number of findings indicate that monomeric actin might not be the only nonfilamentous form of actin in the cell but that unconventional assemblies of actin may exist (Pederson & Aebi, 2002). Antibodies that specifically

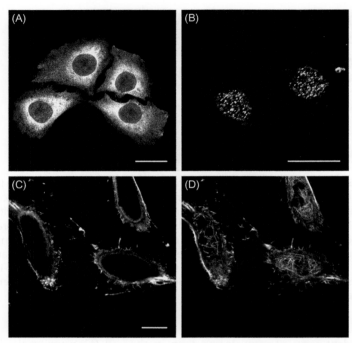

Figure 5.4 Different forms of actin in the nucleus of mammalian cells. (A) Immunofluorescence using the monoclonal antibody 1C7 (Schoenenberger et al., 2005) results in a
fine disperse staining pattern of nonpolymerized actin in the cytoplasm and to a lesser
extent in the nucleus of HeLa cells. (B) Dot-like actin structures are detected in the
nucleus of Rat-2 fibroblasts with the monoclonal antibody 2G2 (Gonsior et al., 1999).
Both images represent confocal sections at the level of maximum nucleus diameter.
(C) Stably transfected HeLa cells were Tet-induced to express HA-tagged actin that is
targeted to the nucleus by an upstream nuclear localization signal (NLS). Nonfilamentous HA-NLS-actin is localized to the periphery of the nucleus. (D) Accumulation
of HA-NLS-actin in the nucleus results in a network of phalloidin-stained F-actin. Note
that F-actin does not extend to the nuclear membrane. Scale bars represent 20 μm. (For
color version of this figure, the reader is referred to the online version of this chapter.)

react with such actin conformations may be well suited for revealing their
presence in cells. With this in mind, a monoclonal antibody called
1C7 was established (Schoenenberger et al., 2005) from mice that were
immunized with cross-linked "lower dimer" (LD), an actin configuration
transiently formed during F-actin polymerization *in vitro* (Millonig, Salvo, &
Aebi, 1988; Steinmetz, Goldie, & Aebi, 1997; Steinmetz, Stiffler, Hoenger,
Bremer, & Aebi, 1997). Epitope mapping revealed that 1C7 recognizes a
sequential epitope that is exposed only in nonpolymerized actin forms.

Consistent with this notion, immunofluorescence using 1C7 results in a fine disperse cytoplasmic staining and a less pronounced signal in the nucleus (Fig. 5.4A), whereas F-actin structures are not stained even after cells have been treated with methanol. However, this antibody does not discriminate between monomeric G-actin and actin in the LD configuration. In addition, the cytoplasmic as well as the nuclear staining patterns are very distinct from those of 2G2 (Fig. 5.4B). Taken together, the different staining patterns observed with these highly specific monoclonal antibodies provide convincing evidence for cellular actin configurations other than the canonical G- or F-actin forms.

3.2. Monitoring actin dynamics in cells

A number of reagents including cytochalasin, latrunculin, and phalloidin are known to modify actin polymerization and are widely used in studies on actin. However, their impact on the dynamics of the actin cytoskeleton cannot be monitored directly. In contrast, transient or stable transfection of cells with actin tagged with different fluorescent proteins like GFP allows the study of the organization and dynamics of the actin cytoskeleton *in vivo* (Ballestrem, Wehrle-Haller, & Imhof, 1998). Nonetheless, adverse effects of GFP–actin fusion proteins on cell spreading, migration, and cell adhesion strength have been reported (Feng, Ning Chen, Vee Sin Lee, Liao, & Chan, 2005).

In recent years, Lifeact fused to fluorescent proteins has found increasing applications as marker for the visualization of F-actin in living cells. Lifeact is a peptide that corresponds to the first 17 residues of the *Saccharomyces cerevisiae* protein Abp140 and binds to actin *in vivo* and *in vitro* (Riedl et al., 2008). Because the binding affinity of the peptide for F-actin is rather low (Riedl et al., 2008), and thus, competition with endogenous ABPs limited, it is generally assumed that Lifeact does not interfere with the dynamics and functionality of actin in cells. Nevertheless, perturbances of the endogenous actin distribution cannot be entirely ruled out. In fact, alteration of the dynamics of actin in cells expressing high levels of Lifeact has been recently reported (Van der Honing, van Bezouwen, Emons, & Ketelaar, 2011). As illustrated in Fig. 5.3B, transfection of Rat-2 fibroblasts with Lifeact-mGFP results in an increase of filopodial extensions that is not seen in untransfected cells. Furthermore, stress-induced nuclear actin rods did not bind Lifeact (Munsie, Caron, Desmond, & Truant, 2009), suggesting that the Abp140 peptide does not interact with all forms of actin.

4. THERE IS MORE THAN F- AND G-ACTIN IN EUKARYOTIC CELLS

As described earlier, different structural forms and arrangements of actin are present in distinct subcellular locations. Some of these conformations are distinct to those present in the mature filament and in the conventional monomeric actin. Thus, they are at risk to be "overlooked" by the classical tools for actin visualization. In fact, unveiling some of these forms in cells has required the development of specific tools (Silván et al., 2012).

4.1. MAL–actin

A fascinating example where actin molecules assume an "unconventional conformation" has been identified in the actin-mediated control of the transcriptional coactivator MAL/MRTF-A (myocardin-related transcription factor A), which transduces Rho GTPase signals to the serum response factor, a transcription factor that controls the activity of many genes in muscle and nonmuscle cells (Cen et al., 2003; Miralles, Posern, Zaromytidou, & Treisman, 2003). Prior to Rho GTPase activation, MAL resides in the cytoplasm, where it interacts with G-actin. Upon activation, Rho GTPases stimulate F-actin polymerization in the cytoplasm, which results in a depletion of G-actin. This, in turn, causes the dissociation of G-actin from MAL that brings about a conformational change. Now in an extended conformation, the bipartite nuclear localization signal (NLS) at the N-terminal RPEL domain of MAL is able to interact with importin α adapter protein and importin-mediated translocation to the nucleus takes place (Miralles et al., 2003; Pawłowski, Rajakylä, Vartiainen, & Treisman, 2010). In addition to inhibiting the nuclear import of MAL, actin may also be involved in regulating its CRM1-mediated export from the nucleus (Vartiainen, Guettler, Larijani, & Treisman, 2007).

The N-terminal actin-binding "RPEL" domain of MAL has three tandem repeats designated RPEL 1–3, where RPEL1 and RPEL2 each form a stable 1:1 complex with G-actin and RPEL3 binds to G-actin much more weakly (Mouilleron, Guettler, Langer, Treisman, & McDonald, 2008). For actin binding, two helices of RPEL contact the hydrophobic cleft of actin between SD1 and SD3 and a hydrophobic stretch in SD3. It is noteworthy that these MAL binding sites partially overlap with the canonical binding sites for profilins or thymosin beta 4. A detailed analysis of the crystal structure of actin–MAL complexes revealed that the orientation and interactions of actin protomers in the RPEL assemblies are not related to those in F-actin (Mouilleron, Langer, Guettler, McDonald, & Treisman, 2011). Thus, it seems unlikely that this actin assembly is able to nucleate F-actin. This is in marked contrast to the multiple actin complexes assembled by tandemly oriented W2 domains found in many F-actin nucleating ABPs (see e.g., Rebowski et al., 2008). In these complexes, actin can be considered an assembly of monomers in a conventional F-actin configuration.

4.2. Actin in the nucleus

The existence of actin in the nucleus was first reported in 1969 (Lane, 1969) and followed by occasional reports describing nuclear actin in various cell types from different organisms (Clark & Rosenbaum, 1979; Jockusch, Becker, Hindennach, & Jockusch, 1974; Lestourgeon, Forer, Yang, Bertram, & Pusch, 1975; Nakayasu & Ueda, 1983; Scheer, Hinssen, Franke, & Jockusch, 1984). However, due to the conspicuous absence of phalloidin staining, nuclear actin met with skepticism for decades, even more so as nuclear preparations were thought to be contaminated with actin from the cytosol. To some extent, these concerns were dispelled by studies with hand-isolated nuclei from *Xenopus laevis* and *Pleurodeles waltii* oocytes where an actin concentration between 70 and 90 µM was reported (Clark & Merriam, 1977; Clark & Rosenbaum, 1979; Gounon & Karsenti, 1981). However, only after the identification of actin in several nuclear complexes, which implicates it in diverse nuclear activities including transcription, chromatin remodeling, and nucleocytoplasmic trafficking (reviewed in De Lanerolle & Serebryannyy, 2011), skepticism has finally given way to the challenging search for specific states and/or unique conformations actin may adopt to fulfill its nuclear functions. With the exception of the giant oocyte nuclei of *X. laevis*, usually referred to as the germinal vesicle (Bohnsack, Stüven, Kuhn, Cordes, & Görlich, 2006), actin does not seem to form large filamentous structures in the nucleus. Instead, nuclear actin is believed to be in nonpolymerized state or forming small oligomers. However, the existence of supramolecular structures cannot be discarded since common detection procedures may overlook unconventional actin species. Consistent with this notion, FRAP analysis of a GFP–actin-expressing cell line has provided evidence of a polymeric actin species in the nucleus of living cells (McDonald, Carrero, Andrin, de Vries, & Hendzel, 2006). Interestingly, cultured cells that were subjected to stress, for example, heat shock or dimethyl sulfoxide treatment, increase the import of actin to the nucleus. The accumulation of actin typically leads to the formation of intranuclear paracrystalline rods (Sanger, Sanger, Kreis, & Jockusch, 1980) that were found to contain cofilin (Fukui, 1978; Nishida et al., 1987). As already pointed out, the cofilin-induced structural modification of the filament prevents the binding of the toxin although cofilin and phalloidin do not share a binding site. Other tools, such as GFP-Lifeact, also fail to detect these structures (Munsie et al., 2009).

Actin antibodies specific for different configurations and arrangements have unveiled the existence of distinct actin populations in the nucleus. For example, the monoclonal antibody 1C7 labels nuclear actin assemblies in addition to the nonfilamentous actin in the cytoplasm (Fig. 5.4A). This pattern is clearly distinct from that of the previously mentioned 2G2 antibody, which labels an actin conformation prominently present in the nucleus (Fig. 5.4B) although in a cell cycle–dependent manner. At present, it is not understood how the specific actin configurations revealed by antibodies relate to different nuclear functions. Moreover, the presumed oligomeric state of actin in the nuclear structures remains an enigma. Several pieces of evidence indicate that at least some nuclear functions of actin, for example, transcription (Miyamoto, Pasque, Jullien, & Gurdon, 2011; Obrdlik & Percipalle, 2011; Scheer et al., 1984; Ye, Zhao, Hoffmann-Rohrer, & Grummt, 2008) and DNA double-strand break repair (Andrin et al., 2012) involve a polymeric form of actin or actin associated with some ill-defined filaments (reviewed in Gieni & Hendzel, 2009). In fact, stably transfected HeLa cells that accumulate NLS-tagged cytoplasmic actin in the nucleus upon tetracycline induction show a phalloidin-stained F-actin network throughout the nucleus surrounded by a rim of nonpolymerized actin at the inner nuclear membrane (Fig. 5.4C and D). These findings corroborate that actin is polymerization competent also in the nucleus but is prevented from assembling conventional F-actin under physiological conditions. It is conceivable that specific interactions with ABPs shape the actin configurations in different nuclear compartments.

5. UNCONVENTIONAL ACTIN CONTACTS

The binding of fluorescent phalloidin to actin filaments specifically marks their F-actin conformation. In contrast, as mentioned earlier, changes in the filament morphology, such as those induced by the binding of cofilin (McGough et al., 1997), may prevent phalloidin binding, and hence, these filamentous actin structures are not visualized. The unusual structural properties of actin isolated from *Toxoplasma gondii* also interfere with a classical phalloidin staining (Sahoo, Beatty, Heuser, Sept, & Sibley, 2006). More generally speaking, supramolecular actin structures other than conventional F-actin may go unnoticed because tools for their detection are lacking. As detailed below, the analysis of contacts in actin crystals has demonstrated a number of interactions between subunits that seem to be incompatible with the structure of mature F-actin. However, to discern between fortuitous crystal contacts and those with a biological significance requires the development of suitable tools.

5.1. Crystalline and paracrystalline arrays of actin

The first evidence for actin–actin contacts that differ from those in F-actin filaments was provided by two-dimensional microcrystals. In the presence of the trivalent lanthanide gadolinium, G-actin assembles into tightly packed two-dimensional crystalline arrays that exhibit structural polymorphism depending on the ionic strength used in the crystallization process (Aebi, Smith, Isenberg, & Pollard, 1980; Dos Remedios & Dickens, 1978). The different types of sheets and tubes observed are built from the same basic actin lattice. Depending on the ionic strength used in crystallization, two basic actin lattices can associate back-to-back, either in a parallel fashion or rotated by about 90° relative to each other, to yield either rectangular or square-type double-layered sheets (Fig. 5.5A; Aebi, Fowler, Isenberg,

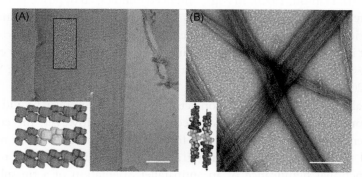

Figure 5.5 Alternative subunit contacts in supramolecular actin structures assembled *in vitro*. (A) In the presence of gadolinium, actin assembles into large 2D crystalline arrays called "sheets". The high magnification inset outlined in black reveals the highly ordered packing of subunits in sheets. As schematically illustrated in the other inset, adjacent subunits are packed in an antiparallel orientation (highlighted in lighter gray) along one axis in the plane of the sheet. (B) At high Mg^{2+} concentrations (>25 mM), actin forms bundles of laterally aligned filaments. The model in the inset illustrates that subunits making contacts between two neighboring filaments with opposite polarity are in an antiparallel orientation. Scale bars represent 200 nm in (A) and 100 nm in (B).

Pollard, & Smith, 1981). Alternatively, the basic lattice can bend and thereby form a tube. The arrangement of actin monomers within the basic crystalline lattice involves antiparallel packing into dimers with p2 symmetry. The actin monomers within the basic lattice are oriented so that the filament axis would run parallel to the sheet plane, and the intersubunit contacts in this direction are similar to those existing along the two long-pitch helical strands of the F–actin filament. The other, antiparallel intersubunit contacts are not found in the mature actin filament. Regularly packed actin assemblies such as sheets, tubes, or ribbons (Steinmetz, Stiffler, et al., 1997) may ultimately yield structural relationships to *in vivo* relevant actin oligomers or polymers such as, for example, the "LD" discussed below or supramolecular assemblies formed thereof.

Furthermore, a number of experimental conditions induce a paracrystalline arrangement of actin filaments. For example, polycations (e.g., polylysine) are known to induce actin polymerization and to cause lateral aggregation of the resulting filaments (Oriol-Audit, 1978). It has been suggested that the polymorphism observed in polylysine-induced F–actin paracrystals is based on differences in the stagger and/or polarity of adjacent filaments in single-layered paracrystals and by superposition of these in multilayered paracrystals (Fowler & Aebi, 1982). Another standard method for inducing actin filament paracrystals has long been the addition of 25–100 mM $MgCl_2$ to G- or F-actin (Fig. 5.5B; Hanson, 1973). In Mg^{2+}-induced bundles, an opposite polarity of the laterally aligned actin filaments has been demonstrated (Francis & DeRosier, 1990). As a result, unconventional actin contacts similar to those present in gadolinium–induced crystalline sheets have been revealed by chemical cross-linking and specific antibody labeling (Millonig et al., 1988; Silván et al., 2012).

Analysis of the intersubunit contacts in crystal structures can provide useful information about unique actin configurations. However, the contact between the molecules can also be fortuitous and therefore need not be related to physiologically occurring actin configurations. In this regard, the functional significance of an antiparallel orientation of actin subunits has remained doubtful for some time.

5.2. The actin "LD"

The formation of the "LD" was first described more than 20 years ago when chemical cross-linking with 1,4-phenylene bismaleimide (PBM) revealed two actin dimers with different electrophoretic mobilities at early stages of polymerization *in vitro* (Millonig et al., 1988). The first dimer to appear, named "LD" for its relative migration on an SDS–polyacrylamide gel, is transient and, with ongoing polymerization, is replaced by the "upper dimer" (UD). While the arrangement of the two actin subunits in the UD corresponds to those in the F–actin filament, subunits in the LD are in an antiparallel orientation, which superficially seems to be incompatible with the helical symmetry of mature F-actin and the subunit interactions defining this geometry. Consistent with this notion, PBM cross-linking of mature actin filaments followed by depolymerization produced only UD and no LD. However, the work of Steinmetz and colleagues showed that there is a correlation between the biochemical detection of LD by PMB cross-linking and the ragged morphology of growing filaments revealed by electron microscopy at early stages of *in vitro* polymerization (Fig. 5.6; Schoenenberger, Bischler, Fahrenkrog, & Aebi, 2002; Steinmetz, Goldie, et al., 1997; Steinmetz, Stiffler, et al., 1997). At subsequent time points, when LD formation decreased and was replaced by UD, filament morphology also smoothed. Ultimately, at steady state, PBM cross-linking reveals only UD and higher oligomers, and corresponding filaments exhibit the

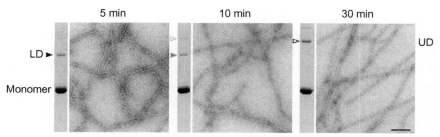

Figure 5.6 Lower dimer formation correlates with a ragged filament morphology in growing F-actin. G-(Ca^{2+})-actin was induced to polymerize by 100 mM KCl, and an aliquot was chemically cross-linked with 1,4-phenylenbismaleimide at the time points indicated. Cross-linked actin was analyzed by SDS-PAGE. In parallel, an aliquot was removed from polymerizing actin and immediately processed for transmission electron microscopy. Five minutes after initiation of polymerization, cross-linked LD (solid arrowheads) is the only dimeric species detected. Concomitant with LD formation, filaments exhibit a ragged morphology. With ongoing polymerization, a second dimeric species called upper dimer (UD; open arrowheads) with a reduced electrophoretic mobility emerges. The morphology of filaments at corresponding time points is significantly smoother although lateral protrusions are still evident at 10 min after salt addition. After 30 min, the filaments have further matured and exhibit a smooth, highly ordered helical structure. Cross-linking of corresponding samples generated UD and higher oligomers. *Image adapted with permission from Steinmetz, Goldie (1997). ©Rockefeller University Press.*

helically ordered, smooth structure typical for mature F-actin. These findings implied that LD transiently incorporated into growing filaments via one of its subunits. If purified actin is polymerized under experimental conditions that sustain LD formation until steady state is reached, the mature filaments also exhibit a ragged morphology. Further support for the propensity of LD to incorporate into growing F-actin is provided by the copolymerization of G-actin and cross-linked LD which results in filaments with laterally protruding moieties that correspond sizewise to an actin subunit (Schoenenberger, Mannherz, & Jockusch, 2011). The association of transient LD incorporation with protruding structures and a higher incidence of branched filaments in the presence of LD led to the notion that this actin configuration might play a role in actin patterning *in vivo*, for example, by facilitating actin filament branching.

Using polylysine to induce actin polymerization and latrunculin A, which prevents the formation of filaments but not that of LD, Bubb et al. (2002) were able to accumulate crystallization-competent LD in the absence of cross-linker and crystallize it. Their LD crystal structure confirmed an antiparallel orientation of the two actin subunits within the LD. Similarly, crystallization of uncomplexed G-actin that had been cleaved by ECP32 protease between Gly42 and Val43 revealed an asymmetric unit containing an antiparallel actin dimer (Klenchin et al., 2006).

Several techniques, such as chemical cross-linking, EM, and fluorescence spectroscopy, support a role for LD in actin plasticity, and yet the existence of LD-mediated actin configurations in cells remained vague. To this end, proteins that induce and stabilize this particular arrangement of actin subunits are further evidence pointing to a physiological role of the LD actin configuration. For example, it has been shown that gelsolin, an ABP with nucleation, filament severing, and capping activity, binds two actin molecules in a spatial configuration similar to that of LD (Hesterkamp, Weeds, & Mannherz, 1993). Moreover, a number of toxins, such as toxofilin and swinholide A (Bubb, Spector, Bershadsky, & Korn, 1995; Lee, Hayes, Rebowski, Tardieux, & Dominguez, 2007), bind two actin subunits in an antiparallel orientation. Toxofilin from *T. gondii* has also been shown to cap actin filaments (Poupel, Boleti, Axisa, Couture-Tosi, & Tardieux, 2000), which is consistent with the notion that filament ends might assume an LD-like configuration.

To examine whether such LD-like actin configurations do exist in cells, we produced and characterized a rabbit polyclonal antibody that specifically recognizes the LD (Silván et al., 2012). Immunofluorescence studies at the

light and electron microscopy level (Fig. 5.7A and B, respectively) documented that an LD configuration is associated with endolysosomal vesicles
(Silván et al., 2012). In addition, fixation of fibroblasts with ice-cold methanol revealed that some distinct actin moiety beneath the plasma membrane
at the leading edge also assumes an LD configuration in cultured mammalian
cells (Fig. 5.7C). Intriguingly, gelsolin, which has been shown to promote
LD-like actin structures *in vitro* (Hesterkamp et al., 1993), was also reported
to mediate the interaction of actin with cell membranes and participate in
vesicle motility (Chen, Murphy-Ullrich, & Wells, 1996; Cunningham,

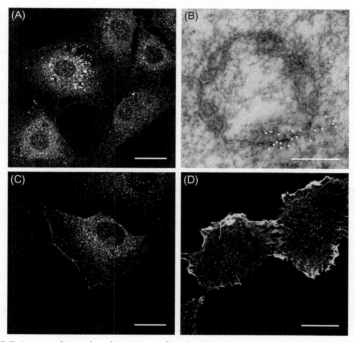

Figure 5.7 Lower dimer localization in fixed cells. (A) Immunofluorescence with the
LD-specific antibody reveals a fine vesicular, mostly cytoplasmic staining pattern in
fibroblasts that partially colocalizes with endolysosomal markers (Silván et al., 2012).
(B) Ultrathin sections of MDCK-II cells that were labeled pre-embedding with gold-
conjugated LD-antibodies (gold dots enlarged for better visibility) confirm that this
unconventional actin configuration is associated with cytoplasmic vesicles. (C) A population of LD assemblies is also detected beneath the plasma membrane in Rat-2 cells
treated with ice-cold methanol. (D) Immunolabeling of Rat-2 cells that were gently
extracted with octyl-POE prior to glutaraldehyde fixation shows that the 2G2-stained
actin configuration is also enriched at the cell periphery. Scale bars represent 20 μm
in A, C and D, and 300 nm in D. (For color version of this figure, the reader is referred
to the online version of this chapter.)

Stossel, & Kwiatkowski, 1991; Hartwig, Chambers, & Stossel, 1989). There-fore, it is tempting to argue that gelsolin triggers the formation of LD con-tacts *in vivo* and stabilizes them at specific cellular sites.

At the molecular level, the effects of LD on actin polymerization are hardly understood. Data on the involvement of LD in nucleation are ambiguous, leaving much room for speculation. The formation of UD from monomeric actin is the thermodynamically most adverse step of F–actin filament formation (Oosawa & Asakura, 1975; Sept & McCammon, 2001; Tobacman & Korn, 1983). Since the LD arrangement is stabilized by two salt bridges located in the dimer interface, it might represent an energetically more favorable polymer-ization intermediate in the nucleation step than the UD (Bubb et al., 2002).

Furthermore, based on crystallographic evidence of actin trimers with two subunits being in a parallel F–actin-like (i.e., UD) conformation and a third one in an LD configuration relative to one of the other two subunits, Reutzel et al. (2004) proposed an LD-mediated filament elongation mech-anism. In their model, each strand of the growing F–actin filament incorpo-rates one of the LD subunits of the trimer. The UDs protruding from each F–actin filament strand are then incorporated into the growing filament through a rotational and translational movement.

In the LD arrangement, the highly conserved C-terminal cysteine resi-due (Cys374) of both actin subunits lies within the dimer interface. Accord-ingly, in the LD crystal structure solved by Bubb et al. (2002), the Cys374 thiol groups of adjacent actin subunits formed a disulfide bond that could aid in the maintenance of these particular arrangement in a redox-dependent manner (Lassing et al., 2007). Interestingly, polymerization of G–actin under oxidizing conditions, which are conducive to the formation of disulfide bridges between antiparallel actin subunits, has been reported to enhance the flexibility of F–actin networks *in vitro* and induce the formation of inter-filament contacts (Tang, Janmey, Stossel, & Ito, 1999).

Correlative chemical cross-linking and ultrastructural morphology stud-ies as well as copolymerization of G–actin and chemically cross-linked LD supported the hypothesis that LD may contribute to the supramolecular patterning of actin, in particular by mediating filament branching and bun-dling (Schoenenberger et al., 2002; Steinmetz, Goldie, et al., 1997; Steinmetz, Stiffler, et al., 1997). Modeling of two adjacent filaments that share an LD by each having incorporated one of its subunits resulted in a branched structure with filaments that are separated by an angle of nearly 30° (Reutzel et al., 2004) with their barbed (i.e., fast growing) ends pointing in opposite directions. According to the dendritic nucleation model, which has emerged as the dominant conceptual picture of the actin organization at the leading edge of migrating cells (Mullins, Heuser, & Pollard, 1998; Pollard, Blanchoin, & Mullins, 2000), the filaments are all oriented with their fast-growing ends toward the plasma membrane. Therefore, it seems rather unlikely that a filament configuration with opposite polarity occurs in the lamellipodium or is associated with the trafficking of cytoplasmic vesicles. To overcome this discrepancy, Reutzel and colleagues proposed major structural changes in the LD contacts that would allow angle variations in the ensuing filament arrangement. In addition, *in vivo*, LD would not be constrained by crystal lattice contacts and both subunits may therefore have free range of rotational movement about the disulfide bond.

An alternative branching model involving two LDs from adjacent filaments has been recently proposed (Silván et al., 2012). In this model, the respective protruding subunit of two LDs interacts with each other to form a parallel UD-like contact. In this manner, both filaments have their barbed ends pointing in the same direction and the branching angle between filaments is approximately 67°. Since this angle is remarkably similar to the 70° angle observed in Arp2/3-mediated branching (Mullins et al., 1998), one might contemplate a relationship between LD- and Arp2/3-mediated side nucleation. Electron microscopy of mildly extracted fibroblasts has revealed the existence of a dendritic F-actin network with similar branch angles at the leading edge of lamellipodia (Svitkina & Borisy, 1999). On the other hand, cryoelectron tomography of *Dictyostelium* cells has revealed the existence of filament junctions with angles ranging from 35° to 90° close to the cell periphery (Medalia et al., 2002). The range of angles observed at filament junctions suggests that filament branching involving one and/or two LDs might occur in the cell. In fact, based on the *in vitro* observation that LD-related branching during F-actin assembly occurs even in the absence of the Arp2/3 complex (Schoenenberger et al., 2002), one might speculate that an LD-like conformation acts as the initial spark in filament branching which triggers the Arp2/3 complex into action. Accordingly, conformational changes in the mother filament were shown to be involved in Arp2/3-mediated branching (Rouiller et al., 2008).

6. CONCLUSION

In this essay, we recapitulate findings that highlight the pronounced structural plasticity of actin, regulated in time and space by numerous ABPs that shape its configuration. Taken together, there is convincing evidence that in eukaryotic cells, there is more to actin than conventional G- and F-actin. Although it is undisputed that these two forms represent the majority of actin structures in cells, other configurations also exist, with largely unknown structural signatures. Because some of the generally used tools to detect actin fail to recognize these unconventional forms, their existence has gone largely unnoticed for too long. A better understanding of these structures could unravel some of the most puzzling questions about actin. In particular, we are eager to learn if unconventional configurations help solve the riddle of form and function of actin in the nucleus.

ACKNOWLEDGMENTS

The authors thank Roman Jakob (Biozentrum, University of Basel) for his help in preparing Fig. 5.1. We also thank Ueli Aebi (M.E. Müller Institute for Structural Biology, University of Basel) for valuable discussions and comments on the manuscript. The work is supported by the Swiss National Science Foundation (grant 31003A_12714 to CAS), the Maurice E. Müller Foundation, and the Deutsche Forschungsgemeinschaft (BMJ).

REFERENCES

Aebi, U., Fowler, W. E., Isenberg, G., Pollard, T. D., & Smith, P. R. (1981). Crystalline actin sheets: Their structure and polymorphism. *The Journal of Cell Biology, 91*, 340–351.

Aebi, U., Millonig, R., Salvo, H., & Engel, A. (1986). The three-dimensional structure of the actin filament revisited. *Annals of the New York Academy of Sciences, 483*, 100–119.

Aebi, U., Smith, P. R., Isenberg, G., & Pollard, T. D. (1980). Structure of crystalline actin sheets. *Nature, 288*, 296–298.

Andrin, C., McDonald, D., Attwood, K. M., Rodrigue, A., Ghosh, S., Mirzayans, R., et al. (2012). A requirement for polymerized actin in DNA double-strand break repair. *Nucleus, 3*, 384–395.

Ballestrem, C., Wehrle-Haller, B., & Imhof, B. A. (1998). Actin dynamics in living mammalian cells. *Journal of Cell Science, 111*, 1649–1658.

Belmont, L. D., Orlova, A., Drubin, D. G., & Egelman, E. H. (1999). A change in actin conformation associated with filament instability after Pi release. *Proceedings of the National Academy of Sciences of the United States of America, 96*, 29–34.

Bohnsack, M. T., Stüven, T., Kuhn, C., Cordes, V. C., & Görlich, D. (2006). A selective block of nuclear actin export stabilizes the giant nuclei of Xenopus oocytes. *Nature Cell Biology, 8*, 257–263.

Bork, P., Sander, C., & Valencia, A. (1992). An ATPase domain common to prokaryotic cell cycle proteins, sugar kinases, actin, and hsp70 heat shock proteins. *Proceedings of the National Academy of Sciences of the United States of America, 89*, 7290–7294.

Bremer, A., & Aebi, U. (1992). The structure of the F-actin filament and the actin molecule. *Current Opinion in Cell Biology, 4*, 20–26.

Bubb, M. R., Govindasamy, L., Yarmola, E. G., Vorobiev, S. M., Almo, S. C., Somasundaram, T., et al. (2002). Polylysine induces an anti-parallel actin dimer that nucleates filament assembly: Crystal structure at 3.5-A resolution. *The Journal of Biological Chemistry, 277*, 20999–21006.

Bubb, M. R., Spector, I., Bershadsky, A. D., & Korn, E. D. (1995). Swinholide A is a microfilament disrupting marine toxin that stabilizes actin dimers and severs actin filaments. *The Journal of Biological Chemistry, 270*, 3463–3466.

Cen, B., Selvaraj, A., Burgess, R. C., Hitzler, J. K., Ma, Z., Morris, S. W., et al. (2003). Megakaryoblastic leukemia 1, a potent transcriptional coactivator for serum response factor (SRF), is required for serum induction of SRF target genes. *Molecular and Cellular Biology, 23,* 6597–6608.

Chen, P., Murphy-Ullrich, J. E., & Wells, A. (1996). A role for gelsolin in actuating epidermal growth factor receptor-mediated cell motility. *The Journal of Cell Biology, 134,* 689–698.

Chik, J. K., Lindberg, U., & Schutt, C. E. (1996). The structure of an open state of beta-actin at 2.65 A resolution. *Journal of Molecular Biology, 263,* 607–623.

Clark, T. G., & Merriam, R. W. (1977). Diffusible and bound actin nuclei of Xenopus laevis oocytes. *Cell, 12,* 883–891.

Clark, T. G., & Rosenbaum, J. L. (1979). An actin filament matrix in hand-isolated nuclei of X. laevis oocytes. *Cell, 18,* 1101–1108.

Cooper, J. A. (1987). Effects of cytochalasin and phalloidin on actin. *The Journal of Cell Biology, 105,* 1473–1478.

Cunningham, C. C., Stossel, T. P., & Kwiatkowski, D. J. (1991). Enhanced motility in NIH 3T3 fibroblasts that overexpress gelsolin. *Science, 251,* 1233–1236.

Dawson, J. F., Sablin, E. P., Spudich, J. A., & Fletterick, R. J. (2003). Structure of an F-actin trimer disrupted by gelsolin and implications for the mechanism of severing. *The Journal of Biological Chemistry, 278,* 1229–1238.

De La Cruz, E. M., Mandinova, A., Steinmetz, M. O., Stoffler, D., Aebi, U., & Pollard, T. D. (2000). Polymerization and structure of nucleotide-free actin filaments. *Journal of Molecular Biology, 295,* 517–526.

De Lanerolle, P., & Serebryannyy, L. (2011). Nuclear actin and myosins: Life without filaments. *Nature Cell Biology, 13,* 1282–1288.

Dos Remedios, C. G., & Dickens, M. J. (1978). Actin microcrystals and tubes formed in the presence of gadolinium ions. *Nature, 276,* 731–733.

Fan, J., Saunders, M. G., & Voth, G. A. (2012). Coarse-graining provides insights on the essential nature of heterogeneity in actin filaments. *Biophysical Journal, 103,* 1334–1342.

Feng, Z., Ning Chen, W., Vee Sin Lee, P., Liao, K., & Chan, V. (2005). The influence of GFP-actin expression on the adhesion dynamics of HepG2 cells on a model extracellular matrix. *Biomaterials, 26,* 5348–5358.

Fowler, W. E., & Aebi, U. (1982). Polymorphism of actin paracrystals induced by polylysine. *The Journal of Cell Biology, 93,* 452–458.

Francis, N. R., & DeRosier, D. J. (1990). A polymorphism peculiar to bipolar actin bundles. *Biophysical Journal, 58,* 771–776.

Fujii, T., Iwane, A. H., Yanagida, T., & Namba, K. (2010). Direct visualization of secondary structures of F-actin by electron cryomicroscopy. *Nature, 467,* 724–728.

Fukui, Y. (1978). Intranuclear actin bundles induced by dimethyl sulfoxide in interphase nucleus of Dictyostelium. *The Journal of Cell Biology, 76,* 146–157.

Galkin, V. E., Orlova, A., Schröder, G. F., & Egelman, E. H. (2010). Structural polymorphism in F-actin. *Nature Structural and Molecular Biology, 17,* 1318–1323.

Gallwitz, D., & Seidel, R. (1980). Molecular cloning of the actin gene from yeast Saccharomyces cerevisiae. *Nucleic Acids Research, 8,* 1043–1059.

Gieni, R. S., & Hendzel, M. J. (2009). Actin dynamics and functions in the interphase nucleus: Moving toward an understanding of nuclear polymeric actin. *Biochemical Cell Biology, 87,* 283–306.

Gimona, M., Vandekerckhove, J., Goethals, M., Herzog, M., Lando, Z., & Small, J. V. (1994). Beta-actin specific monoclonal antibody. *Cell Motility and the Cytoskeleton, 27,* 108–116.

Gonsior, S. M., Platz, S., Buchmeier, S., Scheer, U., Jockusch, B. M., & Hinssen, H. (1999). Conformational difference between nuclear and cytoplasmic actin as detected by a monoclonal antibody. *Journal of Cell Science, 112,* 797–809.

Gounon, P., & Karsenti, E. (1981). Involvement of contractile proteins in the changes in consistency of oocyte nucleoplasm of the newt Pleurodeles waltlii. *The Journal of Cell Biology, 88*, 410–421.

Graceffa, P., & Dominguez, R. (2003). Crystal structure of monomeric actin in the ATP state. Structural basis of nucleotide-dependent actin dynamics. *The Journal of Biological Chemistry, 278*, 34172–34180.

Hanson, J. (1967). Axial period of actin filaments. *Nature, 213*, 353–356.

Hanson, J. (1973). Evidence from electron microscope studies on actin paracrystals concerning the origin of the cross-striation in the thin filaments of vertebrate skeletal muscle. *Proceedings of the Royal Society of London Series B: Biological Sciences, 183*, 39–58.

Hartwig, J. H., Chambers, K. A., & Stossel, T. P. (1989). Association of gelsolin with actin filaments and cell membranes of macrophages and platelets. *The Journal of Cell Biology, 108*, 467–479.

Hesterkamp, T., Weeds, A. G., & Mannherz, H. G. (1993). The actin monomers in the ternary gelsolin: 2 actin complex are in an anti-parallel orientation. *European Journal of Biochemistry, 218*, 507–513.

Holmes, K. C., Popp, D., Gebhard, W., & Kabsch, W. (1990). Atomic model of the actin filament. *Nature, 347*, 44–49.

Holmes, K. C., Sander, C., & Valencia, A. (1993). A new ATP-binding fold in actin, hexo-kinase and Hsc70. *Trends in Cell Biology, 3*, 53–59.

Jockusch, B. M., Becker, M., Hindennach, I., & Jockusch, H. (1974). Slime mould actin: Homology to vertebrate actin and presence in the nucleus. *Experimental Cell Research, 89*, 241–246.

Jones, L. J., Carballido-López, R., & Errington, J. (2001). Control of cell shape in bacteria: Helical, actin-like filaments in Bacillus subtilis. *Cell, 104*, 913–922.

Kabsch, W., & Holmes, K. C. (1995). The actin fold. *The FASEB Journal, 9*, 167–174.

Kabsch, W., Mannherz, H. G., Suck, D., Pai, E. F., & Holmes, K. C. (1990). Atomic structure of the actin:DNase I complex. *Nature, 347*, 37–44.

Kindle, K. L., & Firtel, R. A. (1978). Identification and analysis of Dictyostelium actin genes, a family of moderately repeated genes. *Cell, 15*, 763–778.

Klenchin, V. A., Khaitlina, S. Y., & Rayment, I. (2006). Crystal structure of polymerization-competent actin. *Journal of Molecular Biology, 362*, 140–150.

Lane, N. J. (1969). Intranuclear fibrillar bodies in actinomycin D-treated oocytes. *The Journal of Cell Biology, 40*, 286–291.

Lassing, I., Schmitzberger, F., Björnstedt, M., Holmgren, A., Nordlund, P., Schutt, C. E., et al. (2007). Molecular and structural basis for redox regulation of beta-actin. *Journal of Molecular Biology, 370*, 331–348.

Lazarides, E., & Weber, K. (1974). Actin antibody: The specific visualization of actin filaments in non-muscle cells. *Proceedings of the National Academy of Sciences of the United States of America, 71*, 2268–2272.

Lee, S. H., Hayes, D. B., Rebowski, G., Tardieux, I., & Dominguez, R. (2007). Toxofilin from Toxoplasma gondii forms a ternary complex with an anti-parallel actin dimer. *Proceedings of the National Academy of Sciences of the United States of America, 104*, 16122–16127.

Lestourgeon, W. M., Forer, A., Yang, Y. Z., Bertram, J. S., & Pusch, H. P. (1975). Contractile proteins. Major components of nuclear and chromosome non-histone proteins. *Biochimica et Biophysica Acta, 379*, 529–552.

Lorenz, M., Popp, D., & Holmes, K. C. (1993). Refinement of the F-actin model against X-ray fiber diffraction data by the use of a directed mutation algorithm. *Journal of Molecular Biology, 234*, 826–836.

McDonald, D., Carrero, G., Andrin, C., de Vries, G., & Hendzel, M. J. (2006). Nucleoplasmic beta-actin exists in a dynamic equilibrium between low-mobility polymeric species and rapidly diffusing populations. *The Journal of Cell Biology, 172*, 541–552.

McGough, A., & Chiu, W. (1999). ADF/cofilin weakens lateral contacts in the actin filament. *Journal of Molecular Biology*, *291*, 513–519.

McGough, A., Pope, B., Chiu, W., & Weeds, A. (1997). Cofilin changes the twist of F-actin:
Implications for actin filament dynamics and cellular function. *The Journal of Cell Biology*,
138, 771–781.

Medalia, O., Weber, I., Frangakis, A. S., Nicastro, D., Gerisch, G., & Baumeister, W. (2002).
Macromolecular architecture in eukaryotic cells visualized by cryoelectron tomography.
Science, *298*, 1209–1213.

Millonig, R., Salvo, H., & Aebi, U. (1988). Probing actin polymerization by intermolecular
cross-linking. *The Journal of Cell Biology*, *106*, 785–796.

Minehardt, T. J., Kollman, P. A., Cooke, R., & Pate, E. (2006). The open nucleotide pocket
of the profilin/actin X-ray structure is unstable and closes in the absence of profilin.
Biophysical Journal, *90*, 2445–2449.

Miralles, F., Posern, G., Zaromytidou, A.-I., & Treisman, R. (2003). Actin dynamics control
SRF activity by regulation of its coactivator MAL. *Cell*, *113*, 329–342.

Miwa, T., Manabe, Y., Kurokawa, K., Kamada, S., Kanda, N., Bruns, G., et al. (1991). Structure, chromosome location, and expression of the human smooth muscle (enteric type)
gamma-actin gene: Evolution of six human actin genes. *Molecular and Cellular Biology*, *11*,
3296–3306.

Miyamoto, K., Pasque, V., Jullien, J., & Gurdon, J. B. (2011). Nuclear actin polymerization is
required for transcriptional reprogramming of Oct4 by oocytes. *Genes & Development*,
25, 946–958.

Morton, W. M., Ayscough, K. R., & McLaughlin, P. J. (2000). Latrunculin alters the actin-
monomer subunit interface to prevent polymerization. *Nature Cell Biology*, *2*, 376–378.

Mouilleron, S., Guettler, S., Langer, C. A., Treisman, R., & McDonald, N. Q. (2008).
Molecular basis for G-actin binding to RPEL motifs from the serum response factor
coactivator MAL. *The EMBO Journal*, *27*, 3198–3208.

Mouilleron, S., Langer, C. A., Guettler, S., McDonald, N. Q., & Treisman, R. (2011).
Structure of a pentavalent G-actin*MRTF-A complex reveals how G-actin controls
nucleocytoplasmic shuttling of a transcriptional coactivator. *Science Signaling*, *4*, ra40.

Mullins, R. D., Heuser, J. A., & Pollard, T. D. (1998). The interaction of Arp2/3 complex
with actin: Nucleation, high affinity pointed end capping, and formation of branching
networks of filaments. *Proceedings of the National Academy of Sciences of the United States of
America*, *95*, 6181–6186.

Munsie, L. N., Caron, N., Desmond, C. R., & Truant, R. (2009). Lifeact cannot visualize
some forms of stress-induced twisted F-actin. *Nature Methods*, *6*, 317.

Nakayasu, H., & Ueda, K. (1983). Association of actin with the nuclear matrix from bovine
lymphocytes. *Experimental Cell Research*, *143*, 55–62.

Nishida, E., Iida, K., Yonezawa, N., Koyasu, S., Yahara, I., & Sakai, H. (1987). Cofilin
is a component of intranuclear and cytoplasmic actin rods induced in cultured cells.
Proceedings of the National Academy of Sciences of the United States of America, *84*,
5262–5266.

Obrdlik, A., & Percipalle, P. (2011). The F-actin severing protein cofilin-1 is required for
RNA polymerase II transcription elongation. *Nucleus*, *2*, 72–79.

Oda, T., Iwasa, M., Aihara, T., Maéda, Y., & Narita, A. (2009). The nature of the globular-
to fibrous-actin transition. *Nature*, *457*, 441–445.

Oda, T., Namba, K., & Maéda, Y. (2005). Position and orientation of phalloidin in F-actin
determined by X-ray fiber diffraction analysis. *Biophysical Journal*, *88*, 2727–2736.

Oosawa, F., & Asakura, S. (1975). *Thermodynamics of the polymerization of protein*. London and
New York: Academic Press.

Oriol-Audit, C. (1978). Polyamine-induced actin polymerization. *European Journal of
Biochemistry*, *87*, 371–376.

Otterbein, L. R., Graceffa, P., & Dominguez, R. (2001). The crystal structure of uncomplexed actin in the ADP state. *Science*, *293*, 708–711.

Oztug Durer, Z. A., Diraviyam, K., Sept, D., Kudryashov, D. S., & Reisler, E. (2010). F-actin structure destabilization and DNase I binding loop: Fluctuations mutational cross-linking and electron microscopy analysis of loop states and effects on F-actin. *Journal of Molecular Biology*, *395*, 544–557.

Pawłowski, R., Rajakylä, E. K., Vartiainen, M. K., & Treisman, R. (2010). An actin-regulated importin α/β-dependent extended bipartite NLS directs nuclear import of MRTF-A. *The EMBO Journal*, *29*, 3448–3458.

Pederson, T., & Aebi, U. (2002). Actin in the nucleus: What form and what for? *Journal of Structural Biology*, *140*, 3–9.

Pollard, T. D., Blanchoin, L., & Mullins, R. D. (2000). Molecular mechanisms controlling actin filament dynamics in nonmuscle cells. *Annual Review of Biophysics and Biomolecular Structure*, *29*, 545–576.

Popp, D., Narita, A., Lee, L. J., Ghoshdastider, U., Xue, B., Srinivasan, R., et al. (2012). Novel actin-like filament structure from Clostridium tetani. *The Journal of Biological Chemistry*, *287*, 21121–21129.

Popp, D., & Robinson, R. C. (2011). Many ways to build an actin filament. *Molecular Microbiology*, *80*, 300–308.

Poupel, O., Boleti, H., Axisa, S., Couture-Tosi, E., & Tardieux, I. (2000). Toxofilin, a novel actin-binding protein from Toxoplasma gondii, sequesters actin monomers and caps actin filaments. *Molecular Biology of the Cell*, *11*, 355–368.

Rebowski, G., Boczkowska, M., Hayes, D. B., Guo, L., Irving, T. C., & Dominguez, R. (2008). X-ray scattering study of actin polymerization nuclei assembled by tandem W domains. *Proceedings of the National Academy of Sciences of the United States of America*, *105*, 10785–10790.

Reutzel, R., Yoshioka, C., Govindasamy, L., Yarmola, E. G., Agbandje-McKenna, M., Bubb, M. R., et al. (2004). Actin crystal dynamics: Structural implications for F-actin nucleation, polymerization, and branching mediated by the anti-parallel dimer. *Journal of Structural Biology*, *146*, 291–301.

Riedl, J., Crevenna, A. H., Kessenbrock, K., Yu, J. H., Neukirchen, D., Bista, M., et al. (2008). Lifeact: A versatile marker to visualize F-actin. *Nature Methods*, *5*, 605–607.

Rouiller, I., Xu, X. P., Amman, K. J., Egile, C., Nickell, S., Nicastro, D., et al. (2008). The structural basis of actin filament branching by the Arp2/3 complex. *The Journal of Cell Biology*, *180*, 888–895.

Rould, M. A., Wan, Q., Joel, P. B., Lowey, S., & Trybus, K. M. (2006). Crystal structures of expressed non-polymerizable monomeric actin in the ADP and ATP states. *The Journal of Biological Chemistry*, *281*, 31909–31919.

Sahoo, N., Beatty, W., Heuser, J., Sept, D., & Sibley, L. D. (2006). Unusual kinetic and structural properties control rapid assembly and turnover of actin in the parasite Toxoplasma gondii. *Molecular Biology of the Cell*, *17*, 895–906.

Sanger, J. W., Sanger, J. M., Kreis, T. E., & Jockusch, B. M. (1980). Reversible translocation of cytoplasmic actin into the nucleus caused by dimethyl sulfoxide. *Proceedings of the National Academy of Sciences of the United States of America*, *77*, 5268–5272.

Saunders, M. G., & Voth, G. A. (2012). Comparison between actin filament models: Coarse-graining reveals essential differences. *Structure*, *20*, 641–653.

Sawaya, M. R., Kudryashov, D. S., Pashkov, I., Adisetiyo, H., Reisler, E., & Yeates, T. O. (2008). Multiple crystal structures of actin dimers and their implications for interactions in the actin filament. *Acta Crystallographica Section D: Biological Crystallography*, *64*, 454–465.

Scheer, U., Hinssen, H., Franke, W. W., & Jockusch, B. M. (1984). Microinjection of actin-binding proteins and actin antibodies demonstrates involvement of nuclear actin in transcription of lampbrush chromosomes. *Cell*, *39*, 111–122.

Schoenenberger, C.-A., Bischler, N., Fahrenkrog, B., & Aebi, U. (2002). Actin's propensity for dynamic filament patterning. *FEBS Letters*, *529*, 27–33.

Schoenenberger, C.-A., Buchmeier, S., Boerries, M., Sütterlin, R., Aebi, U., & Jockusch, B. M. (2005). Conformation-specific antibodies reveal distinct actin structures in the nucleus and the cytoplasm. *Journal of Structural Biology*, *152*, 157–168.

Schoenenberger, C.-A., Mannherz, H. G., & Jockusch, B. M. (2011). Actin: From structural plasticity to functional diversity. *European Journal of Cell Biology*, *90*, 797–804.

Schroeder, U., Graff, A., Buchmeier, S., Rigler, P., Silvan, U., Tropel, D., et al. (2009). Peptide nanoparticles serve as a powerful platform for the immunogenic display of poorly antigenic actin determinants. *Journal of Molecular Biology*, *386*, 1368–1381.

Scoville, D., Stamm, J. D., Toledo-Warshaviak, D., Altenbach, C., Phillips, M., Shvetsov, A., et al. (2006). Hydrophobic loop dynamics and actin filament stability. *Biochemistry*, *45*, 13576–13584.

Sept, D., & McCammon, J. A. (2001). Thermodynamics and kinetics of actin filament nucleation. *Biophysical Journal*, *81*, 667–674.

Silván, U., Boiteux, C., Sütterlin, R., Schroeder, U., Mannherz, H. G., Jockusch, B. M., et al. (2012). An anti-parallel actin dimer is associated with the endocytic pathway in mammalian cells. *Journal of Structural Biology*, *177*, 70–80.

Skalli, O., Ropraz, P., Trzeciak, A., Benzonana, G., Gillessen, D., & Gabbiani, G. (1986). A monoclonal antibody against alpha-smooth muscle actin: A new probe for smooth muscle differentiation. *The Journal of Cell Biology*, *103*, 2787–2796.

Steinmetz, M. O., Goldie, K. N., & Aebi, U. (1997). A correlative analysis of actin filament assembly, structure, and dynamics. *The Journal of Cell Biology*, *138*, 559–574.

Steinmetz, M. O., Stiffler, D., Hoenger, A., Bremer, A., & Aebi, U. (1997). Actin: From cell biology to atomic detail. *Journal of Structural Biology*, *119*, 295–320.

Steinmetz, M. O., Stoffler, D., Müller, S. A., Jahn, W., Wolpensinger, B., Goldie, K. N., et al. (1998). Evaluating atomic models of F-actin with an undecagold-tagged phalloidin derivative. *Journal of Molecular Biology*, *276*, 1–6.

Straub, F. B. (1942). Actin. *Stud. Inst. Med. Chem. Univ. Szeged.*, *2*, 3–15.

Straub, F. B., & Feuer, G. (1950). Adenosine triphosphate, the functional group of actin. *Kísérletes Orvostudomány*, *2*, 141–151.

Svitkina, T. M., & Borisy, G. G. (1999). Progress in protrusion: The tell-tale scar. *Trends in Cell Biology*, *24*, 432–436.

Tang, J. X., Janmey, P. A., Stossel, T. P., & Ito, T. (1999). Thiol oxidation of actin produces dimers that enhance the elasticity of the F-actin network. *Biophysical Journal*, *76*, 2208–2215.

Tobacman, L. S., & Korn, E. D. (1983). The kinetics of actin nucleation and polymerization. *The Journal of Biological Chemistry*, *258*, 3207–3214.

Van den Ent, F., Amos, L. A., & Löwe, J. (2001). Prokaryotic origin of the actin cytoskeleton. *Nature*, *413*, 39–44.

Van der Honing, H. S., van Bezouwen, L. S., Emons, A. M. C., & Ketelaar, T. (2011). High expression of Lifeact in Arabidopsis thaliana reduces dynamic reorganization of actin filaments but does not affect plant development. *Cytoskeleton (Hoboken)*, *68*, 578–587.

Vartiainen, M. K., Guettler, S., Larijani, B., & Treisman, R. (2007). Nuclear actin regulates dynamic subcellular localization and activity of the SRF cofactor MAL. *Science*, *316*, 1749–1752.

Wulf, E., Deboben, A., Bautz, F. A., Faulstich, H., & Wieland, T. (1979). Fluorescent phallotoxin, a tool for the visualization of cellular actin. *Proceedings of the National Academy of Sciences of the United States of America*, *76*, 4498–4502.

Ye, J., Zhao, J., Hoffmann-Rohrer, U., & Grummt, I. (2008). Nuclear myosin I acts in concert with polymeric actin to drive RNA polymerase I transcription. *Genes & Development*, *22*, 322–330.

CHAPTER SIX

Chromatin Reorganization Through Mitosis

Paola Vagnarelli[1]

Heinz Wolff Building, Brunel University, Uxbridge, United Kingdom
[1]Corresponding author: e-mail address: Paola.Vagnarelli@brunel.ac.uk

Contents

Abstract

Chromosome condensation is one of the major chromatin-remodeling events that occur during cell division. The changes in chromatin compaction and higher-order structure organization are essential requisites for ensuring a faithful transmission of the replicated genome to daughter cells. Although the observation of mitotic chromosome condensation has fascinated Scientists for a century, we are still far away from understanding how the process works from a molecular point of view.

In this chapter, I will analyze our current understanding of chromatin condensation during mitosis with particular attention to the major molecular players that trigger and maintain this particular chromatin conformation. However, within the chromosome, not all regions of the chromatin are organized in the same manner. I will address separately the structure and functions of particular chromatin domains such as the centromere. Finally, the transition of the chromatin through mitosis represents just an interlude for gene expression between two cell cycles. How the transcriptional information that governs cell linage identity is transmitted from mother to daughter represents a big and interesting question. I will present how cells take care of the aspect ensuring that mitotic chromosome condensation and the block of transcription does not wipe out the cell identity.

Advances in Protein Chemistry and Structural Biology, Volume 90
ISSN 1876-1623
http://dx.doi.org/10.1016/B978-0-12-410523-2.00006-7

179

1. PACKAGING THE GENOME

1.1. The interphase nucleus

The genetic material itself consists of linear molecules of deoxyribonucleic acid (DNA) that, in each human diploid cell, correspond to approximately 6 million base pairs (bp) generating a total length about 2 m. This DNA is enclosed within a nucleus of about 10 nm in diameter. Clearly the packing of the DNA must be a very controlled process that deals with a few intuitive challenges. The first is the neutralization of the negative charges of, in particular, the phosphate molecules to allow the DNA to bend and compact within a very small volume; the second is the organization that allows a regulated execution of the main vital functions such as replication, transcription, and repair of DNA damage. At the first level of packaging, there are protein particles called nucleosomes and around them the 147 bp of DNA is wound 1.75 times. This constitutes the minimal unit of chromatin (Blobel et al., 2009; Kornberg & Thomas, 1974; reviewed by Kornberg & Lorch, 1999a, 1999b). Each nucleosome is composed of an octamer of highly positively charged proteins called "Histones": two each of histones H2A, H2B, H3, and H4. This first level of organization provides both the initial compaction and the important neutralization of the negative charges of the DNA. Each single unit follows along the DNA molecule, and they are separated by a stretch of 10–50 bp of naked DNA. This arrangement constitutes the so-called beads on a string or 10-nm fiber (Fig. 6.1).

In living cells, nucleosome positions are in part regulated by intrinsic DNA sequence preferences due to the thermodynamic energetics associated with wrapping stiff DNA around the histone octamer (Drew & Travers, 1985a, 1985b; Jiang & Pugh, 2009a, 2009b; Radman-Livaja & Rando, 2010). In addition, other proteins such as the ATP-dependent chromatin-remodeling complexes can contribute to the positioning. Recent data also shows that some aspects of the transcriptional machinery are critical for this positioning (Hughes, Jin, Rando, & Struhl, 2012).

The N-terminal tails of the histones are disordered (and therefore smeared out in the crystal structure), which give them the potential to bind to many different proteins, a characteristic of intrinsically disordered protein domains. However, the N-terminal tails include many sites where posttranslational modifications such as acetylation, methylation, and phosphorylation can occur. These modifications may modulate histone–DNA interactions, but they also form a "histone code" that mediates different proteins to local

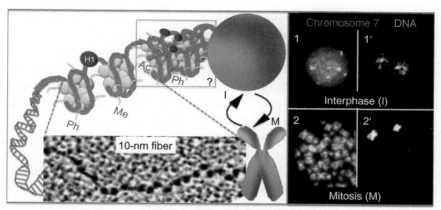

Figure 6.1 Schematic representation of the organization of DNA into the 10-nm fiber: the first step of chromatin compaction. The histone tails within the nucleosome structure (green) can be posttranslational modified with phosphate groups (Ph), methyl groups (Me), or acetyl groups (Ac). How the higher levels of compactions are achieved is still debatable as indicated by the question mark in the gray box. The 10-nm fiber can be visualized as "beads on a string" by electron microscopy as shown in the EM image below the drawing. EM image kindly provided by Prof. W. C. Earnshaw, Edinburgh. The transition from interphase (I) to mitosis (M) is characterized by a remarkable change in the organization of the chromatin (compare 1 and 2) but the effective level of condensation is only two- to threefold variation between the two stages. Compare the area occupied by chromosome 7 (chromosome painting by FISH) in the interphase nucleus (1–1′) versus the area occupied in the mitotic cell (2–2′). (1–2) Chromosome painting with a chromosome 7-specific library of a human lymphoma cell line. FISH images kindly provided by Dr. Sabrina Tosi (Brunel). (See Color Insert.)

chromatin regions. The charges present on the histone tails together with the recruitment of particular proteins contribute to another level of chromatin compaction and this will be addressed later in the chapter.

The second level of compaction is the organization of the 10–nm fiber into the 30-nm fiber. Experiments have shown that this level of compaction depends on the linker histone H1 (Thoma, Koller, & Klug, 1979). It is also accepted that this degree of compaction requires the presence of more positive ions that are provided by the surrounding aqueous solution. In fact, if the chromatin is exposed to a solvent of low ionic strength the fibers unfold into the "beads on strings" of the 10-nm fiber and increasing the salt concentration again can reverse this process. How is the 10-nm fiber folded into this higher-order conformation? Finch and Klug proposed one model in 1976 (Finch & Klug, 1976; Finch et al., 1977). In their model called the "solenoid," consecutive nucleosomes are located next to each other in the fiber, folding into a simple one-start helix. Later, on the basis of microscopic observations of isolated nucleosomes, a second model was proposed (Woodcock, Frado, & Rattner, 1984). This suggested the nucleosomes to be arranged in a zigzag manner. More recently (Robinson, Fairall, Huynh, & Rhodes, 2006; Robinson & Rhodes, 2006), measurements further defined the dimensions of the "30-nm" chromatin fiber: evidence for a compact, interdigitated structure. A third model was proposed according to which the 30-nm fiber is an interdigitated solenoid (Daban & Bermudez, 1998). In addition, another recent study using magnetic tweezers to probe the mechanical properties of long nucleosome arrays has suggested that a one-start helix comprises the topology underlying the 30-nm fiber (single-molecule force spectroscopy reveals a highly compliant helical folding for the 30-nm chromatin fiber; Kruithof et al., 2009). All this data has been obtained from reconstituted chromatin *in vitro*. The question now is does the 30-nm chromatin fiber exists *in vivo*? The answer to the question has been very difficult due to technical reasons; both electrospectroscopy imaging (ESI) and conventional transmission electron microscopy use the projection of data into a single image plane therefore individual chromatin fibers cannot be resolved. Recently, by combining ESI with electron tomography, this previous limitation could be overcome. The study has clearly shown that chromatin domains within the nucleus of mouse cells were exclusively configured as 10-nm chromatin fibers (Fussner et al., 2012). These results are in agreement with data obtained by whole-genome analysis using three-C molecular biology methodologies that show that *in vivo* chromatin is best modeled by a 10-nm chromatin fiber arrangement (Dekker, 2008; Lieberman-Aiden et al., 2009).

At this point, although still debatable, the second level of compaction could also be represented not only by the 30-nm fiber but also by the multiple bending of the basic 10-nm fiber.

2. THE MITOTIC CHROMOSOME

When the cell approaches the phase of segregating the genetic material (cell division or mitosis) into two daughter cells, clearly changes in the organization of the chromatin has to occur. In organisms where the process is accompanied by the removal of the nuclear membrane (open mitosis), the chromatin is subjected to a remodeling process that has as an endpoint the mitotic chromosome. Mitotic chromosomes have fascinated those in the biological research since the nineteenth century. It was in 1880 that

W. Waldeyer first described the chromosomes as we know them today. The use of a particular dye allowed their visualization under the light microscope and the terminology of the chromosome was adopted to indicate the colored (chromo) bodies (soma) that appeared evident during cell division. However, after so many years we still do not know how the interphase chromatin reorganizes in mitosis to give form to the mitotic chromosomes.

The process of mitotic chromosome formation is remarkable and visually very impressive to follow *in vivo*. This dramatic change has always been referred to as "chromosome condensation" giving the impression that from interphase chromatin to mitotic chromosomes a great deal of compaction activity is required. However, if we analyze in detail the difference in compaction between the two stages, we will be quite surprised in discovering that the degree of compaction is much less than it appears. Therefore, most of the changes are due to a different organization of the chromatin structure and possibly the process should be more precisely referred to as "chromosome morphogenesis" rather than "chromosome condensation." A few studies have attempted to analyze the chromatin volume from interphase to mitosis and from mitosis to interphase. Some analyses made use of 3D live imaging of cells where the DNA was marked with florescent histones (typically H2B). Another approach used the FLIM/FRET technique to measure the interaction between H2B histones on the chromatin (Lleres, James, Swift, Norman, & Lamond, 2009; Vagnarelli, 2012). The emerging picture is that the changes in chromatin volume (GFP:H2B) or compaction (FLIM/FRET of H2B) are only two to three times between early prophase to mitosis or between mitosis to G1 (Fig. 6.1: 1 and 2). For example, Martin and Cardoso (2010) analyzed the dynamic behavior of chromatin during the transition from late anaphase to G1 in HeLa. They identified a biphasic process with a first rapid and global decondensation step by a factor of 2, followed by a slower phase in which only part of the chromatin decondensed an additional twofold again.

A further chromatin compaction in mitosis, although minor if compared to the other levels of DNA condensation into chromatin, still requires the neutralization of additional negative charges to allow the process to occur and be maintained during the execution of mitosis (about 1 h for a human cell actively proliferating with a cell cycle of 24 h). Some interesting observations have shown that a major influx of divalent cations (Ca^{2+}, Mg^{2+}) has been evident in the first stages of mitosis (Strick, Strissel, Gavrilov, & Levi-Setti, 2001) and, in particular, the cations Ca^{2+}, Mg^{2+}, Na^+, and K^+ are pivotal to complete and maintain "maximal chromosome condensation" during mitosis. These cations besides providing DNA electrostatic neutralization also seem to be involved in the regulation of a subset of nonhistone proteins functions as well (see below).

The chromosome structure is quite remarkable: each single species has a chromosome set (karyotype) that is unique to the species but also is identical in each mitotic cell. The shape and the length of a particular chromosome are therefore very well controlled. Moreover, if mitotic chromosomes are treated with particular dyes or with trypsin and then stained with Giemsa, substructures within the chromosomes become apparent: these are known as chromosome bands and again they represent a very distinct but conserved feature of the mitotic chromosome. All this suggests that the packaging of the chromatin within the chromosome cannot be solely a stochastic process. Therefore, a very well-controlled mechanism must be in place for both organizing the chromosome structure upon entry into mitosis and for dismantling it after division has occurred.

The question then becomes how is the chromatin organized into the mitotic chromosomes?

To address this question, Laemmli and coworkers in 1978 removed histones from metaphase chromosomes and then observed the results by electron microscopy. In such preparations, the DNA was unfolded in long loops (30–100 kb long) that were attached to an electron-dense region that maintained the shape of the metaphase chromosome and consisted of nonhistone proteins. This network of proteins was defined as the chromosome scaffold (Marsden & Laemmli, 1979), and this observation led the authors to propose the radial loop model for chromosome organization (Laemmli, 1978; Laemmli et al., 1978; Paulson & Laemmli, 1977). At the base of the loop, special DNA sequences (scaffold/matrix-associated region DNA sequences; Razin, 1996) would be attached to the scaffold. A variation of the basic radial loop model is the radial loop/helical coil model; in this case, the prophase chromatid, organized as a radial loop, is then helically folded to form the final metaphase chromosome (Boy de la Tour & Laemmli, 1988; Rattner & Lin, 1985). Common to both models is the key role of nonhistone proteins in organizing the chromosome shape. Alternative models referred to as hierarchical models of chromosome folding postulate that the 10- and 30-nm chromatin fiber folds progressively into larger fibers that coil to form the final metaphase chromosomes (Belmont & Bruce, 1994; Belmont, Sedat, & Agard, 1987; Sedat & Manuelidis, 1978; Zatsepina, Poliakov, & Chentsov Iu, 1983). This model would be less dependent on the formation of a core protein scaffold. Successive helical coiling and folded chromonema models are examples of this group of models (Belmont & Bruce, 1994; Sedat & Manuelidis, 1978). However, biophysical experiments based on mechanical stretching of isolated mitotic chromosomes have shown that chromosomes can be extensively stretched without altering their diameter and or a treatment with nucleases leads to a loss of metaphase chromosome elasticity; these studies argue against a role for scaffold proteins being responsible for the chromosomal mechanical properties (Poirier, Eroglu, Chatenay, & Marko, 2000; Poirier, Eroglu, & Marko, 2002; Poirier & Marko, 2002a). Therefore, an alternative model based on chromatin fibers cross-linked by stabilizing proteins has been proposed (Poirier & Marko, 2002a). The detailed analyses of the early stages of prophase chromosome condensation have allowed Belmont and coworkers to postulate another hypothesis which can reconcile all the previous views on mitotic chromosome structure: "the hierarchical folding, axial glue" model. In this model, the initial folding of large-scale chromatin fibers into early prophase chromosomes releases nuclear-chromatin attachments, and it is followed by a condensation into the uniform, tight middle prophase chromatids of 200–300 nm in diameter. A second coiling of this middle prophase chromatid during late prophase through prometaphase yields the final metaphase chromosome and a network of cross-linked chromatin by scaffold proteins may still serve to provide chromosome stability.

More recently the use of cryo electron microscopy and small angle X-ray scattering (Fussner, Ching, & Bazett-Jones, 2011; Fussner, Djuric, et al., 2011; Maeshima, Hihara, & Takata, 2010; Nishino et al., 2012) have called into question the continuous folding model of chromatin and chromosome condensation. These studies failed in identifying the existence of the 30-nm fiber within the mitotic chromosomes but instead they could only detect the presence of folded 10-nm fibers. These chromatin fibers are then packed in a fractal manner or polymer melt. The fractal nature of the mitotic chromosome could be explained by self-associating mechanisms that are directed by interaction between histone tails. The modification of these parameters can modulate the level of association and therefore account for chromatin states more or less compacted (Hansen, 2002; Liu, Tao, Chen, & Cao, 2012; Szerlong, Prenni, Nyborg, & Hansen, 2010). Even in these latest revised models, the importance of the nonhistone protein network is clear and therefore understanding of what it is and which is its biological functions is clearly an extremely important question in the field of chromosome biology.

2.1. Chromosome scaffold

From the early studies of Marsden and Laemmli (1979), it was quite clear that nonhistone proteins were present in the chromosomes and that they were somehow involved in the assembly or maintenance of the characteristic structure that is a mitotic chromosome. Laemmli and colleagues were the first to isolate purified mitotic chromosomes, remove the DNA and the majority of other chromosome proteins: this procedure led to the isolation of an insoluble protein fraction called the scaffold (Adolph, Cheng, & Laemmli, 1977; Lewis & Laemmli, 1982; Mirkovitch, Mirault, & Laemmli, 1984; Fig. 6.3: 1). The importance of this insoluble fraction from both a biochemical and a biological point of view has been questioned for many years since its discovery. Within the fraction, topoisomerase II (Topo II) (Earnshaw, Halligan, Cooke, Heck, & Liu, 1985; Earnshaw & Heck, 1985; Gasser, Laroche, Falquet, Boy de la Tour, & Laemmli, 1986) and SMC2 (a subunit of the condensin complex) were identified. However, the argument was that this chromosomal fraction does not represent a network of protein that associates with chromatin in mitosis (Earnshaw & Laemmli, 1983; Paulson & Laemmli, 1977) but rather it represents a precipitate that forms with the specific extraction procedure due to insoluble proteins that stick together when there is no DNA.

However, additional evidence down the line suggested that these scaffold components were somehow of importance for the mitotic chromosome structure. For example, Topo II and SMC2 were present in discrete foci on swollen mitotic chromosomes (Earnshaw & Heck, 1985; Saitoh & Laemmli, 1994). Immunostaining of unextracted chromosomes and *in vivo* observations have confirmed the existence of an axial core distribution in native metaphase chromosomes for proteins that are identified within the scaffold fraction: Topo II (Earnshaw & Heck, 1985; Maeshima & Laemmli, 2003; Ono et al., 2003; Tavormina et al., 2002), condensins (Maeshima & Laemmli, 2003; Ono et al., 2003; Tavormina et al., 2002), and KIF4A (Mazumdar, Sundareshan, & Misteli, 2004; Samejima et al., 2012) (Fig. 6.2: 2).

However, the biological significance and the existence of a protein network were not very clear. Only later in 2003, Hudson, Vagnarelli, Gassmann, and Earnshaw (2003) provided the first demonstration for the existence and function of a chromosome scaffold. The study analyzed the cytological and biochemical structure of chromosomes that were lacking one of the major scaffold component: SMC2 (condensin). In the absence

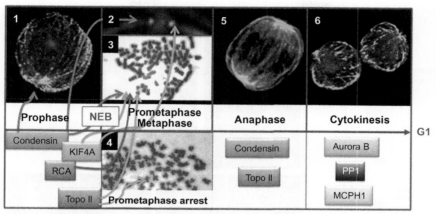

Figure 6.2 Chromatin reorganization during mitosis requires several sequential steps that are regulated and coordinated by different chromosome-associated or regulatory proteins. 1: prophase cell (red: DNA, green: tubulin); 2: metaphase cells. A chromosome (red) is attached to the spindle microtubules (white) via the kinetochore (green). 3: Chromosome in prometaphase assume the characteristic x-shape structure and the sister chromatids are becoming visible. 4: Cell blocked in prometaphase by a spindle poison drug. The chromosomes maintain their mitotic structure while they undergo a further reduction in length. Note that the separation between sister chromatids is more pronounced than in (3) but they are still joined at the primary constriction. 4: Anaphase cell: the sister chromatids (red) are migrating toward the spindle poles and maintain their rod-shape structure. (green: tubulin). 6: Telophase-cytokinesis: the chromosomes reach their maximum compaction in mitosis. See text for explanations on the role of different proteins. (See Color Insert.)

of SMC2, the biochemical scaffold fraction was absent. This observation clearly argued against a scaffold being just a precipitate of insoluble proteins but rather representing a protein network. In this condition, the chromosome morphology was altered with mitotic chromosomes being larger than normal (less compacted) and mainly their structure was very fragile (Fig. 6.2: 3–6). This represented a new parameter in the description of a mitotic chromosome. When chromosomes are prepared for cytological observations, they are usually subjected to a low–salt treatment (hypotonic) that swells the cell and increases the size of a chromosomes thus allowing the high resolution of its substructures (such as primary and secondary constrictions, chromosome banding patters, etc.), but it maintains the overall mitotic chromosome shape. If this procedure is accomplished with chromosomes lacking the condensin complex, then the final product is a chromosome with no shape where the DNA is not well packed and organized (Fig. 6.3: 3–4).

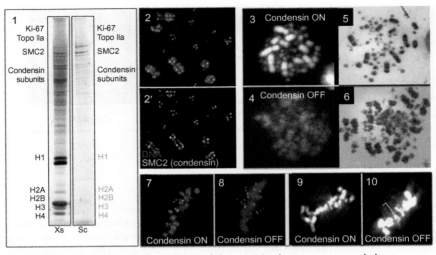

Figure 6.3 1: Biochemical composition of the mitotic chromosomes and chromosome scaffold. Proteins from highly purified chromosomes were separated on a SDS-PAGE gel and silver stained (Xs). A fraction of the chromosome preparation was treated with high salt to remove the histone fraction, micrococcal nuclease to remove the DNA to isolate the scaffold components (Sc). In the fraction, there is enrichment for topoisomerase II (Topo II) and the condensin complex (SMC2 + condensin subunits); preparation and images kindly provided by Dr. Ohta (Japan). 2–2': mitotic chromosomes stained with an antibody against the SCM2 subunit of the condensin complex. SMC2 (green) is found in the middle axis of each sister chromatid. 3–4: Mitotic chromosomes of normal cells (with condensin-condensin ON) (3) and cell where condensin has been removed (condensin OFF) (4). Note the difference in the appearance (shape) of the two preparations. 5–6: Chromosomes from normal cells (5) and cells without condensin (6) were treated with trypsin to produce a chromosome banding (GTG). While in normal cells this binding pattern is recognizable, in chromosomes without condensin there is no clear distinction between light- and dark-stained bands within a chromosome. 7–10: Metaphase from normal cells (7, 9) and cells without condensin (8, 10). In these preparations, the mitotic spindle (not visible in the pictures) is preserved and all the chromosomes are attached to the spindle microtubules (as seen in Fig. 6.2: 2). In cells without condensin, the chromosomes are less robust, the centromeric chromatin is weakened, and it is possible to visualize stretches of chromatin (white bar) emanating from the chromosome masses in the middle of the metaphase plate (10). In 7 and 8, the kinetochores are stained with green and are found quite a distance away from the main chromosome bodies in cells without condensin (8). (For interpretation of the references to color in this figure legend, the reader is referred to the online version of this chapter.)

These observations allowed the development of the first assay for probing the intrinsic chromosome structure (ICS assay) of a mitotic chromosome. This assay is based on two properties of the mitotic chromosomes: (1) if a chromosome is treated with very low salt, the chromatin unfolds up to the bead-on-a-string level (Fig. 6.4: 1–6); however, the addition of divalent

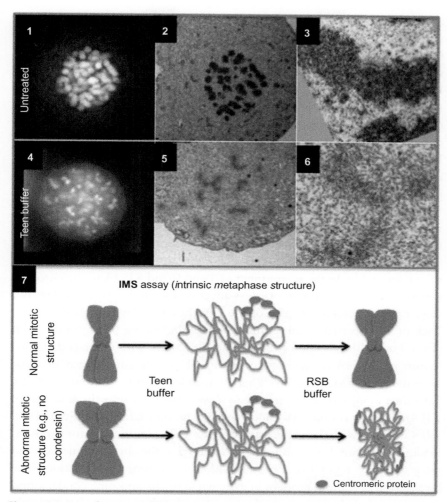

Figure 6.4 Assay for testing the intrinsic metaphase structure (IMS). 1–5; Correlative LM/ EM of normal prometaphase cells (1–3) and cells treated with the TEEN buffer (2–6). 1–2: LM images of the chromosomes stained with DAPI. 3–4: EM of the same cells where in (3) the chromosome are easily recognized and the chromatin is highly compacted (magnification in 5), while in TEEN-treated cells (4) only a shadow of the chromosome is visible and the chromatin is decondensed (down to the 10-nm fiber); 5 and 6 are magnifications of 3 and 4, respectively. LM/EM images kindly provided by D. Booth, Edinburgh. 7: Scheme of the IMS assay and outcome for chromosomes with a normal chromosome structure and for chromosomes with a compromised structure (e.g., lack of condensin or KIF4). (For color version of this figure, the reader is referred to the online version of this chapter.)

cations allows the chromatin to refold and the chromosome regains its original shape and all its properties including band patterns. This procedure can be conducted several times in sequence and at the end we have a chromosome morphologically undistinguishable from the one we have started with. If we start with a chromosome lacking condensin, however, we will never regain the original shape (Fig. 6.4: 7). That condensin was important for chromosome morphogenesis had already been suggested by experiments conducted in a cell-free extract prepared from *Xenopus* eggs (Hirano, Kobayashi, & Hirano, 1997) and genetic analyses have demonstrated an *in vivo* role for condensin subunits in chromosome organization and segregation (Swedlow & Hirano, 2003). The study of Hudson and Vagnarelli (Hudson et al., 2003) has clearly shown that the role of condensin is to organize the mitotic protein network known as scaffold and that the scaffold is essential for the maintenance of the mitotic chromosome structure.

Condensin appears to be at the base of the scaffold assembly; in fact, lack of condensin blocks the recruitment of Topo II and KIF4A to the chromosome axes, but lack of Topo II does affect condensin loading and does not impair the structural integrity of the chromosomes (Spence et al., 2007). Lack of KIF4 partially compromises condensin loading but again does not affect Topo II localization (Samejima et al., 2012). This suggests that condensin and KIF4A are working on a parallel/convergent pathway in chromosome morphogenesis, but Topo II is independent. A recent study by Samejima et al. (2012) has started to shed some light into the function of KIF4A. Using a conditional knock-out system, they have demonstrated that KIF4 interacts with the condensin complex and it is important for part of its loading. How are those components recruited to the chromosome axes and what is their temporal relationship with chromosome condensation? The localization of scaffold proteins in early stages of chromosome condensation remains unclear. In one report, the axial core distribution was observed only for Topo II and not condensins in prophase chromosomes (Maeshima & Laemmli, 2003). This led to the suggestion of a two-step model of chromosome condensation in which Topo II is more central to early stages of chromosome condensation and organization with condensins functioning later. However, functional analyses have indicated that prophase chromosome condensation is delayed in chicken DT40 cells in which the *SMC2* gene is knocked out conditionally (Hudson et al., 2003) or in *Caenorhabditis elegans* embryos depleted of SMC4 (Hagstrom, Holmes, Cozzarelli, & Meyer, 2002), suggesting a role for condensins early in chromosome condensation (Figs. 6.2 and 6.4: 1–6). Belmont and colleagues (Kireeva,

Lakonishok, Kireev, Hirano, & Belmont, 2004) by studying the structural transitions underlying prophase mitotic chromosome condensation and correlating these transitions with the dynamic redistribution of Topo IIα and the condensin SMC2 subunit were able to show that folding of large-scale chromatin fibers is a prominent feature of these early stages of condensation. Large-scale chromatin fibers during early prophase form condensed, linear chromosomes of uniform width in middle prophase that precedes the formation of a well-defined axial core of either SMC2 or Topo II. SMC2 and Topo IIα staining first appears in foci distributed throughout the chromosome width or even at the chromosome exterior. A further doubling of the chromatid diameter and the appearance of a well-defined, central axis of Topo IIα and condensin SMC2 staining occurs well after formation of uniformly condensed chromosomes and a defined chromosome axis in middle prophase. The temporally coordinated appearance of axial staining for both of these proteins down the chromatid center appears as a relatively late event in prophase chromosome condensation.

However, the name of scaffold is a bit misfortunate; it gives the impression that we are dealing with a network that is organized in a very rigid and stiff manner. This is not the case. A few studies have now analyzed the dynamic behavior of some major scaffold proteins, and the message that comes from all of these studies is that several scaffold proteins are actually very dynamic in relation to their presence on the chromosome structure. The most dynamic scaffold component is KIF4A ($t_{1/2}$ 2.5 s) (Samejima et al., 2012) followed by Topo II ($t_{1/2}$ 15 s) (Tavormina et al., 2002). Condensins do still exchange but at much lower rate ($t_{1/2}$ 2 min). These dynamic behaviors could relate to the biology of the mitotic chromosomes that continually need to change to adjust to the mechanical stresses that occur during the alignment of the chromosomes to the mitotic spindle. During this process, the chromokinesins Kid (kinesin-10) and Kif4A (kinesin-4) oppositely tune the polar ejection force and spatially confine chromosomes via position-dependent regulation of kinetochore tension and centromere switch; Kif18A (kinesin-8) attenuates centromere movement by directly promoting microtubule pausing providing the dominant mechanism for restricting centromere movement to the spindle midzone (Stumpff, Wagenbach, Franck, Asbury, & Wordeman, 2012).

The findings that a kinesin (KIF4A) is both a chromosome scaffold protein and also it is involved in chromosome dynamic is quite interesting, and this dual function needs to be addressed in the future to understand how much of the motor activity of this kinesin is necessary for chromosome biogenesis. Therefore, even if we cannot rename this chromosome fraction, we have to at least keep in mind that it represents a dynamic network, and it functions to allow the normal physical properties that a mitotic chromosome requires for the completion of faithful division of the replicated genome.

2.2. What is a mitotic chromosome made of and what does make chromatin condense?

We have identified in the previous section how important is a subset of scaffold protein for the establishment of the mitotic chromosome structure, but it is also quite well established now that chromatin condensation can occur without a chromosome scaffold (Hudson et al., 2003). Are there other candidates that could be responsible for chromosome condensation? Chromosome compaction occurs in prophase before NEB. These early events seem to be triggered by the activation of CyclinB1–Cdk1. Immediately upon activation, a substantial proportion of cyclin B/Cdk1 moves into the nucleus, followed by visible signs of chromosome condensation (Gavet & Pines, 2010a, 2010b); phosphorylation (activation) of condensin appears to be an important step in these early transition stages, and condensins themselves were identified as cyclin/cdk substrates (Kimura, Hirano, Kobayashi, & Hirano, 1998). However, although these phosphorylations seem to be important for the function of condensin, they do not represent the chromatin condensation factor. A chromatin condensation activity named RCA (regulator of chromosome architecture) was discovered to act in early mitosis and being sustained by CDK phosphorylation. This factor is then dephosphorylated by Repo-Man/PP1 complex during anaphase. This chromatin condensation pathway acts in parallel with condensin to build the mitotic chromosome. In fact, even in the absence of compaction, a degree of chromosome condensation can be achieved *in vivo* and this can be maintained even after anaphase providing cyclin B/Cdk1 is maintained active (Vagnarelli et al., 2006). The nature of RCA is still unknown, and the quest for it is still open. Being a condensation factor, it is expected to be quite abundant on chromosomes therefore knowing the detailed composition of a system is a very good starting point toward the understanding of how the system works. This allows us to take it apart piece by piece and identify each critical function of each tile of the puzzle. A great advance in this direction has been made by Earnshaw and colleagues. Ohta et al. isolated highly purified mitotic chromosomes, and these preparations were analyzed by mass spectrometry to identify not only the proteins present in the preparation but also their relative abundance. Using these analyses, they were able to show that 40% of mitotic chromosome protein consists of ~4000 nonhistone proteins (Ohta et al., 2010) (Fig. 6.3: 1). Combining this quantitative proteomics with bioinformatic analysis, they generated a series of independent classifiers that describe the 4000 proteins of the mitotic chromosome proteome. This integrated approach allowed us also to uncover functional relationships between protein complexes in the context of intact chromosomes and has made predictions on the possible allocation of the ~560 uncharacterized proteins identified in the study. The specific contribution of each of these proteins (known and novel) toward mitotic chromosome organization and function will be quite a task to undertake for future research. Whether any of these new components is RCA will be revealed in the coming years.

2.3. Physical properties of the mitotic chromosomes

Measurement of chromosome mechanics provides a powerful tool for analysis of chromosome organization. These analyses have been conducted using isolated newt or human mitotic chromosomes. Chromosomes were shown to have reversible elasticity, with force depending nearly linearly on extension for extensions smaller than 100%. This elasticity is lost by cutting the DNA molecules demonstrating that chromatin itself is the main load-bearing element inside a mitotic chromosome. Digestion of proteins causes a reduction in the force constant but does not change the elasticity of the chromosomes. These studies indicate that the scaffold proteins may represent the cross-linkers for a chromatin network (Bai et al., 2011; Sun, Kawamura, & Marko, 2011).

3. SPECIALIZED MITOTIC CHROMATIN

The mitotic chromosome is a specialized type of chromatin organization *per se*, but within this structure, there are regions where a particular chromatin environment is required to sustain specialized functions. In this section, I will consider, in particular, the centromeric chromatin.

The centromere is the chromosome region where sister chromatids remain joined together until the onset of anaphase, but it is also the locus where the specialized structure that mediates chromosome segregation (the kinetochore) is assembled. The centromere is therefore an extremely important region for cell division and for the control of its error-free execution. The chromatin of the centromere can be divided into two main regions: the inner centromeric chromatin and the inner kinetochore chromatin. I will discuss separately the organization of these two types of chromatin and their significance in chromosome function.

3.1. Inner centromeric chromatin

The inner centromere is the region of chromatin contained between the kinetochore regions. For organisms that have a regional centromere, it is the area where the primary constriction is located. In this discussion, I will not consider the details of the DNA sequences that compose a centromere in vertebrates but only the chromatin organization, its characteristics, and the biological relevance for chromosome segregation. For review on centromere, see Verdaasdonk and Bloom (2011) and Vagnarelli, Ribeiro, and Earnshaw (2008).

This chromosome region is important for maintaining the chromatids together until anaphase onset and also it represents an essential component of the system of forces that control chromosome congression. In this process, a balance between pushing and pulling forces is the key regulatory mechanism. Spindle microtubules attached to kinetochores exert a pulling action in opposite direction of the sister chromatids. This is counteracted by the organization of the centromeric chromatin. Intuitively it can be envisaged that some molecular constrains hold the chromatin together and in turn they prevent that just the pulling forces from the microtubules unravel the chromatin fibers (Fig. 6.2: 2).

The question that now arises is which are the molecular mediators of the resistance forces? One important player is clearly the complex that physically holds the sister chromatids together: the cohesion complex. The cohesin complex consists of two SMC proteins, SMC1 and SMC3, and two non-SMC subunits SCC1 and SCC3. Higher eukaryotes contain two orthologues of SCC3: SA1 and SA2 and two cohesin complexes can be formed with either the SA1 or SA2 subunit, and they are removed from the chromosomes in a two–step process. Most of the arm cohesin complexes are removed in prophase by PLK1–dependent phosphorylation of their SA2 subunit (Hauf et al., 2005; Sumara et al., 2002; Waizenegger, Hauf, Meinke, & Peters, 2000; Warren et al., 2000), while the small fraction of centromeric cohesin is removed at anaphase onset by the cleavage of RAD21 (Hauf, Waizenegger, & Peters, 2001; Uhlmann, Wernic, Poupart, Koonin, & Nasmyth, 2000). Therefore, the fraction of the centromeric cohesin complex is protected until anaphase. In mammals, the two proteins SGOL1 (SGO stands for shugoshin or guardian spirit) and SGOL2 localizes to this chromatin region in mitosis (Kitajima, Kawashima, & Watanabe, 2004).

SGO1 has been shown to be indispensable to protect centromeric cohesion during early mitosis. The other mammalian shugoshin, SGOL2, is not directly involved in the protection of centromeric cohesion during mitosis. Hs-SGOL2 is dispensable for keeping sister chromatids together until the onset of anaphase in HeLa cells, and SGOL2-depleted mitotic cells do not manifest defects in chromatid cohesion (Huang et al., 2007; Orth et al., 2011). The inner centromeric region is also very heterochromatic where the heterochromatin proteins are localized. It is known that HP1 is strongly enriched at the inner centromere (Sugimoto, Tasaka, & Dotsu, 2001). Recent studies have suggested that heterochromatin influences chromosome segregation. In mammals, HP1 associating at centromeric heterochromatin is largely displaced at mitosis, but a small proportion remains at the centromere. The heterochromatin structure at the centromere is required for cohesin retention at metaphase (Fukagawa et al., 2004; Guenatri, Bailly, Maison, & Almouzni, 2004). In human cell, SGO interacts with HP1 alpha and the depletion of HP1 alpha in mitosis causes a reduction of SGO1 at centromere and a premature loss of cohesion suggesting that shugoshin recruitment might be a primary role for centromeric heterochromatin in eukaryotic chromosomes (Yamagishi, Sakuno, Shimura, & Watanabe, 2008). Is this phenomenon related to specific mitotic posttranslational modification of histone H3?

HP1 recognizes Lys9-methylated histone H3, which specifically exists in heterochromatin, and recruits several regulatory proteins (Grewal & Jia, 2007). HP1 contains both a chromodomain (CD) and a chromoshadow domain (CSD) (Koch, Kueng, Ruckenbauer, Wendt, & Peters, 2008; Nielsen et al., 2002; Thiru et al., 2004); the CD recognizes Lys9-methylated histone H3, whereas the CSD interacts with PXVXL-containing proteins. Histone methyltransferase Suv39h, which methylates histone H3 at Lys9, is required for the recruitment of HP1 at the inner centromere (Lachner, O'Carroll, Rea, Mechtler, & Jenuwein, 2001). In mitosis, however, due to the activity of Aurora B kinase, histone H3 is also phosphorylated at Ser 10 generating the mitosis-specific epitope H3K9me3S10ph. This can be found in the pericentric heterochromatin of different eukaryotes. As stated earlier, most HP1 dissociates from chromosomes during mitosis, but if phosphorylation of H3 serine 10 is prevented, HP1 remains chromosome-bound throughout mitosis. H3 phosphorylation by Aurora B is therefore part of a "methyl/phos switch" mechanism that displaces HP1 and perhaps other proteins from mitotic heterochromatin (Hirota, Lipp, Toh, & Peters, 2005). This pathway, however, seems not to be involved in the centromeric cohesion protection and suggests that the HP1α region responsible for hSgo1 localization might use the hinge domain of HP1α for centromeric retention rather than the CD. The HP1 localization in fact seems to be dependent on a centromeric protein hMIS14. This inter-action is still not very well understood. It appears that, for successful local-ization of HP1 at the inner centromere in mitosis, the interaction between HP1 and Mis14 needs to occur in interphase. However, at the transition between interphase and mitosis, this interaction is lost and accumulation of HP1 at the centromere occurs and structural changes must take place so as to dissociate hMis14 from HP1 during the assembly of the separate structures, namely, the inner centromere and kinetochore suggesting a "dynamic remodeling" hypothesis in the transition between interphase and mitosis. Because hMis14 is the stable subunit of the Mis12 hetero-tetrameric complex, we presume that hMis14 acts as the subunit of the com-plex during the hypothetical assembly (inner centromere formation requires hMis14, a trident kinetochore protein that specifically recruits HP1 to human chromosomes; Kiyomitsu, Iwasaki, Obuse, & Yanagida, 2010).

From what we have discussed above, it appears that physical constrains of the chromatin could potentially contribute to provide the aspect of rigidity that is essential for the biological function of the inner centromeric region. Is it true and if so which other molecular effectors are important?

We know that that Type II topoisomerases are archetypal nucleic acid-remodeling enzymes that, using ATP as a cofactor, can variously add or remove DNA supercoils and either form or unlink DNA tangles, (Blobel et al., 2009); Topo II is found at the inner centromeres of vertebrate mitotic chromosomes and several studies on vertebrate systems indicate that this enzyme plays a role in shaping mitotic chromatin.

A number of studies have suggested that Topo II might have a role in influencing the organization of the centromere (Rattner, Hendzel, Furbee, Muller, & Bazett-Jones, 1996; Toyoda & Yanagida, 2006), centro-meric cohesion (Bachant, Alcasabas, Blat, Kleckner, & Elledge, 2002; Takahashi, Basu, Bermudez, Hurwitz, & Walter, 2008; Takahashi & Yanagida, 2000), in different model systems. In vertebrates, the lack of Topo II leads to a shortening of the interkinetochore distance when pulling forces are applied in metaphase suggesting a direct role in the generation of the balance of forces (Spence et al., 2007). In this picture, the cohesin complex would impose the chromatin links, while Topo II would work in an opposite direction to resolve the chromatin links and prepare the sister chromatids free of tangles for anaphase segregation. If this model is correct, it could be predicted that the inhibition of Topo II in mitosis would in part compensate for the lack of cohesin at centromere. This hypothesis was tested both in yeast (Dewar, Tanaka, Nasmyth, & Tanaka, 2004) and in vertebrates (Vagnarelli et al., 2004); in both systems, the biorientation defects present in cohesin-depleted cells were partially restored when Topo II was impaired suggesting that the one, if not the only, role of cohesin in facilitating biorientation is to provide the physical connection between sister centromeres necessary to generate tension upon their biorientation.

The *in vivo* observation of chromosomes in mitosis clearly reveals a great deal of oscillatory movements. Chromosomes oscillate back and forth around the metaphase plate (phenomenon known as chromosome oscillation), the kinetochores are pulled toward the pole and then they relax back together (known as kinetochore breathing or interkinetochore stretching), and the kinetochore plates are pulled apart from each other (known as intrakinetochore stretching).

Micromanipulation of mitotic chromosomes reveals that their elasticity is dictated by both DNA and protein components (Poirier & Marko, 2002b). Among the major components of the nonhistone chromosome proteins, we have analyzed the contribution of cohesin and Topo II but what about condensin?

Several studies in yeasts and vertebrates (Gerlich, Hirota, Koch, Peters, & Ellenberg, 2006; Jaqaman et al., 2010; Oliveira, Heidmann, & Sunkel, 2007; Ribeiro et al., 2009; Stephens, Haase, Vicci, Taylor, & Bloom, 2011) have analyzed the dynamic behavior of the centromeric chromatin in mitosis, and they all observed that indeed the condensin complex contributes to the establishment of the inner centromeric chromatin. In fact, depletion of condensin causes an abnormal and much extended stretching of the inner centromeric chromatin when pulling forces are applied (Fig. 6.3: 7–10). These excursions, although of great amplitude, are reversible. This suggests that the elastic behavior of the chromatin is not lost, and this may just reflect the intrinsic elasticity of the DNA/histone composition.

The centromeric chromatin contributes at the end on the generation of the system of forces in mitosis by acting as a spring and can be modeled with the simplest of the assumption as linear spring (the Hookean spring) where the deformation of the spring is linearly proportional to the force applied and depends on the spring stiffness (spring constant). If this model is applied to the experimental data, it can be clearly shown that depletion of the condensin complex halves the value of the spring constant (Ribeiro et al.,

2009; Samoshkin et al., 2009; Uchida et al., 2009). Therefore, the condensin complex sets most of the compliance of the vertebrate centromere.

In a wider study, where all the dynamic parameters of mitotic chromosomes were taken into account, it was quite clearly shown that that baseline oscillation and breathing speeds in late prometaphase and metaphase are set by microtubule depolymerases, whereas oscillation and breathing periods depend on the stiffness of the mechanical linkage between sisters (Jaqaman et al., 2010).

Which is the biological significance of these oscillatory and stretching motions?

The first insights on the importance of generating tension was first demonstrated with the classic experiments by Bruce Nicklas (Nicklas & Koch, 1969) when he showed that there is a tension-sensing mechanism that detects the presence of tension between the sister chromatids. This system monitors if the sister chromatids of a mitotic chromosome are properly attached to spindle microtubules coming from opposite poles in a fashion known as biorientation. When this occurs, tension is generated between the sister chromatids due to the opposing pulling forces of the spindle and the resistance of the centromeric chromatin (as described above). If both sisters attach to the same pole of the spindle (mono-orientation), the two kinetochores are pulled in the same direction, no physical stretching occurs, and no force is generated between them. This "lack of tension" activates a signaling cascade known as spindle assembly checkpoint (SAC) which prevents the cleavage of the linkage between the sister chromatids and hence progression into anaphase. Once all centromeres are under tension, the spindle checkpoint is turned off, the cohesin linkage between the sister chromatids is broken, the sister chromatids migrate to the opposite poles of the spindle, and faithful chromosome segregation is accomplished. In condensin-depleted cells in fact, mitotic progression is delayed due to a prolonged activation of the SAC. This could be interpreted that kinetochores not under tension are more prone to shrink (catastrophe) than to grow. Since the centromeres without condensin are less compliant, tension is reached at greater stretch and higher is the likelihood of the kinetochores to detach from microtubules. This model has been directly tested by very elegant *in vitro* experiments by Sue Biggins and colleagues. In this study, purified individual kinetochore particles were subjected to physiological levels of tension, and it was shown that they could stabilize kinetochore–microtubule attachments directly. This generated tension also inhibited microtubules catastrophes and promoted rescues (Akiyoshi et al., 2010).

The centromeric chromatin is clearly a signaling platform for the regulation of mitotic progression. As we have seen, this is regulated not only by the physical properties of the inner centromeric chromatin but also other signaling molecules must be involved. We have now a good understanding that nonhistone proteins are involved in the organization of the higher-order structure of the inner centromeric chromatin in mitosis but is the chromatin *per se* (DNA and histones) playing any role at all?

During mitosis, when microtubule/kinetochore attachment occurs in an incorrect manner, sophisticated yet not completely understood molecular machinery is responsible for their correction. A key player in this process is Aurora B, a kinase that modulates the strength of kinetochore–MT attachments by differentially phosphorylating components of the kinetochore (Carmena, Ruchaud, & Earnshaw, 2009; Lampson & Cheeseman, 2011). Aurora B associates with three regulatory subunits (inner centromere protein (INCENP), Survivin, and Borealin) in a quaternary protein assembly, the chromosomal passenger complex (CPC) (for review on the CPC, see Carmena & Earnshaw, 2006; Ruchaud, Carmena, & Earnshaw, 2007; Vagnarelli & Earnshaw, 2004). The catalytic core of the complex is Aurora B and its activator; the C-terminal region of INCENP (the so-called IN-box) (Sessa et al., 2005) and the targeting platform consists of Survivin, Borealin, and the N-terminal region of INCENP (Jeyaprakash et al., 2007). This complex is particularly enriched at the inner centromere of mitotic chromosomes, and its accumulation is higher on centromeres of chromosomes that are not properly attached. The localization of the complex specifically at the centromere of mitotic chromosomes was known to depend on the targeting subunit of the complex INCENP and Survivin (Ainsztein, Kandels-Lewis, Mackay, & Earnshaw, 1998; Yue et al., 2008) but the molecular mechanism has been elusive for some time.

In a screening for mitotic histone-binding proteins, Funabiki and colleagues (Kelly, Ghenoiu, et al., 2010) found that the CPC interacts with histones H3/H4 in *Xenopus* cytostatic factor-arrested egg extracts. In particular, it was shown that the complex recognizes specifically the tail of histone H3 when it is phosphorylated at Thr3 and it is mediated by the scaffolding protein Survivin. Interestingly, this modification occurs only in mitosis at the inner centromeric region and it is accomplished by Haspin kinase. This interaction is important for spindle checkpoint signaling at the centromere in human cells (Higgins, 2010; Kelly, Ghenoiu, et al., 2010; Wang et al., 2010, 2011; Yamagishi, Honda, Tanno, & Watanabe, 2010). These results highlight a mechanism whereby the cell cycle control of Haspin activity restricts chromosomal Aurora B localization and activation to M phase through histone modification.

Detailed analyses of the crystal structure of Survivin bound to the histone H3 tail phosphorylated in Thr3 has shown that Survivin appears to be a rather rigid scaffold both for assembling the core of the CPC (Jeyaprakash et al., 2007) and for docking to nucleosomes via H3-T3ph tails. The mechanism is similar to the one used by PHD fingers. The N-terminal tetrapeptide of histone H3 binds to a zinc finger motif that is a common structural feature of both folds. Affinity and specificity for this recognition come from anchoring the N-terminal Ala residue of H3 in a shallow hydrophobic pocket, whereas fine-tuning by the adjacent pockets permits decoding of the presence or absence of posttranslational modifications on H3-Thr3 and possibly on H3-Lys4 and/or H3-Arg2) (Jeyaprakash et al., 2007). The removal of the Th3 phosphorylation in anaphase allows the CPC to leave the centromere. In fact a sustained Haspin activity even after anaphase onset compromises the CPC transferring to the spindle (Kelly, Ghenoiu, et al., 2010), while a premature removal of this phosphorylation by an hyperactive form of the Repo-Man/PP1 phosphatase causes a displacement of the CPC from the centromere and the inability to correct inappropriate kinetochore—microtubules attachments (Qian, Lesage, Beullens, Van Eynde, & Bollen, 2011; Vagnarelli et al., 2011).

By looking in detail at the distribution of known histone modifications within the inner centromere, it was noted that in prophase and prometaphase chromosomes, H3-pT3 signals and the H2A-pT120 signals are only partially overlapping and Aurora B signals are enriched in the merged region of these signals (Yamagishi et al., 2010).

Depletion of either of these histone modifications would cause a mislocalization of Aurora B. H2A is phosphorylated at thr120 by Bub1 kinase and mediates the recruitment of the CPC via SGO2. Therefore, centromeric localization of the CPC via the Bub1-H2AT120ph-shugoshin-CPC pathway (Liu, Vader, Vromans, Lampson, & Lens, 2009) appears to provide an input signal that boosts the Haspin-H3T3ph-CPC feedback loop specifically at centromeres to drive H3T3ph and CPC concentration at this location. Additionally, because Aurora B influences Bub1 (Dai, Sullivan, & Higgins, 2006; Yamagishi et al., 2008) and Sgo1 localization (Tsukahara, Tanno, & Watanabe, 2010) and centromeric H2A phosphorylation (see above), it is possible that the Bub1-H2AT120ph-shugoshin-CPC pathway constitutes a second feedback loop regulating centromeric CPC. The combination of these feedback loops would ensure robust CPC localization and function at centromeres.

Therefore, the inner centromere in eukaryotes is defined by the intersection of two histone marks, H2A-pS121 and H3-pT3 and constitutes the molecular network centered on shugoshin-cohesin-CPC, which couples sister chromatid cohesion to chromosome biorientation in eukaryotes.

3.2. The inner kinetochore chromatin

At the onset of mitosis, a subset of the centromeric chromatin directs the assembly and the organization of a multicomplex structure known as the kinetochore (for review on kinetochores, see Hori & Fukagawa, 2012; Santaguida & Musacchio, 2009). The kinetochore is essential for proper cell division since it mediates the attachment to the spindle microtubules, the correction of improper attachments and the signaling checkpoint for the transition into anaphase. This region of the centromeric chromatin is characterized by the presence of a histone H3 variant: CENP-A. CENP-A is found at all functional centromeres including the particular class of neocentromeres (centromeres derived from a chromosome region that does not contain the typical centromeric DNA) (see Marshall, Chueh, Wong, & Choo, 2008 for a review on neocentromeres). CENP-A is essential for kinetochore formation (Goshima, Kiyomitsu, Yoda, & Yanagida, 2003; Howman et al., 2000; Oegema, Desai, Rybina, Kirkham, & Hyman, 2001; Regnier et al., 2005).

Nucleosomes containing the centromere-specific histone H3 variant CENP-A provide an important mark to establish a centromere-specific chromatin structure (Black & Cleveland, 2011). Although CENP-A is essential in every organism and it is necessary for kinetochore specification, it cannot be considered just sufficient for its assembly. In fact, ectopic incorporation of CENP-A in vertebrates does not necessary cause kinetochore assembly (Gascoigne et al., 2011). Several other kinetochore proteins have been identified in the past decade and, most of them, depend of the presence of CENP-A for the assembly.

Among them, the CENP-T-W complex represents a very important platform in the kinetochore assembly pipeline. CENP-T-W contains histone-fold domains required to associate with DNA and target to kinetochores (Hori, Okada, Maenaka, & Fukagawa, 2008; Hori, Amano, et al., 2008); it also bears a long N-terminal flexible region that binds directly to outer kinetochore proteins and could represent a physical bridge between the chromatin and the proteinaceous structure (Gascoigne et al., 2011).

Recruitment of CENP-T is sufficient to induce kinetochore formation suggesting that, although downstream of CENP-A, it can at least partially bypass the requirement for CENP-A nucleosomes. Interestingly, CENP-T does not bind to CENP-A nucleosomes (Hori, Okada, et al., 2008) and, in superresolution experiments on extended chromatin fibers, CENP-A and CENP-T do not localize in the same centromeric chromatin regions (Ribeiro et al., 2010).

CENP-T-W binds a second group of histone-fold-containing proteins, the CENP-S-X complex (Amano et al., 2009), and their association generates a stable CENP-T-W-S-X heterotetramer with structural similarity to canonical nucleosomes. This tetrameric complex can supercoil DNA and can form a nucleosome-like structure. The DNA-binding surface of histone H3–H4 in noncentromeric chromatin faces the minor grove of DNA to bend DNA into a superhelix (Luger & Richmond, 1998a, 1998b). This DNA-binding surface is conserved in the CENP-S-X-T-W complexes, and mutations of these residues results in the reduction of the supercoiling activities *in vitro* and functional kinetochore formation *in vivo*. All these suggest that this complex is integral part of chromatin structure that is at the basis for the kinetochore maturation.

It is known that within the centromeric chromatin not all nucleosomes contain CENP-A and, in fact, the pattern of micrococcal nuclease digestion from CENP-T-associated chromatin is different from that of DNA isolated by CENP-A immunoprecipitations (Hori, Amano, et al., 2008). It appears therefore to be at least an alternate pattern between CENP-T-containing nucleosomes and CENP-A containing nucleosomes (as can be visualized by superresolution). The CENP-A-containing chromatin is therefore a very structured region containing repeats of different chromatin flavors: CENP-A domains alternate with canonical H3 blocks and possibly CENP-T-W blocks within kinetochore chromatin (Hori, Amano, et al., 2008; Ribeiro et al., 2010) although these latter have not been mapped at superresolution level together with other H3 modifications. However, the modification pattern of the canonical histone blocks seems to be a less conserved characteristic. In human and *Drosophila* interphase prekinetochores, H3K4me2, but not H3K9me3, was found to be intercalated between CENP-A subdomains (Sullivan & Karpen, 2004) and, in a human artificial chromosome H3K4me2 and lower levels of H3K9me3, were readily detected within the centromeric domain (Nakano et al., 2008). In contrast, levels of H3K4me2 were much lower in one human neocentromere (Alonso, Hasson, Cheung, & Warburton, 2010) and in maize centromeres (Jin et al., 2008; Shi & Dawe, 2006).

What these different patterns represent and what is their function in centromere specification is a very important question but more investigations are required to provide an answer.

The CENP-A-containing chromatin domain, although it exists already in interphase, undergoes a change in physical/chemical properties during mitosis.

We mentioned at the beginning that chromatin can be unraveled by exposing it to low salt (exposure of chromosomes to low ionic strength TEEN buffer (a low-salt buffer in which chromatin higher-order structures are destabilized)). Vagnarelli and colleagues have also shown that this assay can unravel the chromatin but it preserves at least part of the kinetochore protein binding (Ribeiro et al., 2009; Fig. 6.4: 7). Therefore, we can apply this assay to study the properties of the centromeric chromatin both in interphase and mitosis. Both interphase prekinetochores and mitotic kinetochores are much more resistant to unfolding in this buffer than pericentromeric heterochromatin (Ribeiro et al., 2009), suggesting that the inner centromeric chromatin and the centromeric chromatin have different organization structures. However, interphase centromeric chromatin (containing CENP-A) can be extended much more easily than the mitotic counterpart: 17-fold extension for interphase and 8-fold extension for mitosis. Combining this assay with a genetic approach, it was found that the centromeric protein CENP-C is the stabilizer for the centromeric chromatin in the transition from interphase to mitosis. How this occurs is not completely clear but it is plausible that the recruitment of further kinetochore proteins in mitosis could provide an additional source for clamping the chromatin and to confer the necessary stiffness to the structure to allow a solid base for interaction with the spindle microtubules (Ribeiro et al., 2010).

These studies based on sequential chromatin unfolding have also allowed proposing a model for the chromatin organization within the centromeric domain: the boustrophedon model (from the Greek word ox–turning). According to this model, alternating CENP-A and H3 domains fold into a planar sinusoidal patch and several patches would be stacked on top of each other. Such a topology would allow kinetochore size to vary according to the number of microtubules bound with minimal perturbation of local packing and CENP-C could represent the molecular clamp that holds the layers together. It could also be that mitotic-specific chromatin modifications favor local chromatin interaction and stabilize the mitotic centromeric chromatin.

In recent years, Henikoff and colleagues by atomic force microscopy and topological analyses of CENP-A chromatin have suggested that CENP-A nucleosomes can adopt a noncanonical structure and be present as "hemisome" (a single copy of H2A/H2B/CENP-A/H4) instead of the classical octameric version (Dalal, Furuyama, Vermaak, & Henikoff, 2007; Dalal, Wang, Lindsay, & Henikoff, 2007). These observations and their interpretation have caused a great deal of discussion in the scientific community and more research has been directed to investigate the existence of this nucleosomal organization. From those studies conducted both in yeast and human, it has emerged that in both systems the CENP-A-containing nucleosomes vary in structure during the cell cycle. In budding yeast, Shivaraju et al. propose that Cse4 (yeast CENP-A) is in a hemisome configuration during G1, S phase, and G2. During anaphase, this structure changes and CENP-A would form an octameric nucleosome (Shivaraju et al., 2012).

In humans, however, CENP-A exists within a hemisome throughout the cell cycle except at the G1/S transition, when CENP-A nucleosomes become octameric (Bui et al., 2012).

How these transitions occur and their biological significance is still big questions for the future.

At the transition from interphase to mitosis, CENP-A is phosphorylated at serine 7 in its N-terminal tail domain. This phosphorylation event is thought to be initiated by Aurora A in prophase but is then maintained by Aurora B until anaphase onset. What is the significance of this phosphorylation is not quite clear. One hypothesis is that it can represent an important step in kinetochore maturation or, in alternative, it is a relict of evolution but maintained due to the similarity between CENP-A and histone H3 and the importance of histone H3Ser10 phosphorylation in chromatin reorganization in mitosis. Mitotic cells in which the phosphorylation of CENP-A on Ser-7 is prevented, exhibit problems in chromosome alignment and a delayed mitotic progression as a result of a defect in the ability of kinetochores to attach to microtubules (Kunitoku et al., 2003). In these conditions, it was also observed that Aurora B fails to accumulate at the inner centromere in prometaphase and cells exhibit abnormalities at the final stage of cytokinesis (Zeitlin, Shelby, & Sullivan, 2001). Nevertheless, the real meaning of this mitotic modification is still very unclear.

4. GENE BOOKMARKING

We have quite extensively discussed how chromatin is reorganized during mitosis and how this remodeling is associated with a further level of chromatin compaction. Another important aspect that is characteristic of mitotic chromatin is that it becomes transcriptional incompetent.

However, after cell division, transcription needs to be resumed in an orderly fashion and also cell linage specificity needs to be perpetuated and maintained. Clearly a mechanism must be in place to identify important genes that need be activated and transcribed in the next cell cycle thereby ensuring a safe passage through mitosis. This phenomenon of maintenance of a specific transcription pattern through generation is known as gene bookmarking. Initially, the term was used to refer to gene promoters that present hypersensitivity to nuclease in mitotic cells (Groudine & Weintraub, 1982; Martinez-Balbas, Dey, Rabindran, Ozato, & Wu, 1995; Sarge & Park-Sarge, 2009); now it is used to indicate a specific retention of transcription factor at a specific locus throughout mitosis.

At the onset of mitosis, transcription factors are either degraded or displaced from the chromatin and ETS1, octane-binding transcription factor 2 (OCT2; also known as POU2F2), B-my and SP1 are specific examples. These factors will then be resynthesized or reassociated with the specific chromatin region in a temporal manner that is typical for a specific cell linage.

General factors, such as members of the BET family, the histone methyltransferase MLL and, albeit variably, the acetyltransferase p300, have been reported to remain bound to subsets of genes during mitosis (Blobel et al., 2009; Dey et al., 2000; Kanno et al., 2004; Kouskouti & Talianidis, 2005; Kruhlak et al., 2001; Zaidi et al., 2003). Moreover, the basal transcription factor TBP (TATA binding protein) has been reported to mark active genes during mitosis in some studies (Chen, Hinkley, Henry, & Huang, 2002; Christova & Oelgeschlager, 2002; Xing, Vanderford, & Sarge, 2008) but not in others (Blobel et al., 2009; Komura, Ikehata, & Ono, 2007; Segil, Guermah, Hoffmann, Roeder, & Heintz, 1996; Varier et al., 2010). Finally, a select few sequence-specific DNA-binding proteins remain at least partially associated with chromatin during mitosis, including RUNX2, AP2, HSF2, and FoxI1 (Martinez-Balbas et al., 1995; Xing et al., 2005; Young, Hassan, Pratap, et al., 2007; Young, Hassan, Yang, et al., 2007; for review, see Egli, Birkhoff, & Eggan, 2008). All these could represent different ways of specifically marking a subset of genes during mitosis.

How does gene bookmarking work? Gene bookmarking appears to come in several flavors, and the bookmarked genes are not regulated by a single epigenetic mechanism. Several new studies have identified key feature of gene bookmarking for a particular transcription factor, and I will surmise some of the advances in the field that had shed some light on this phenomenon. One of the first observations was that a number of gene promoters, including those of the hsp70i (inducible heat shock protein 70) and c-myc genes, do not appear to be tightly compacted in mitotic cells (Duncan et al., 1994; Levens et al., 1997; Michelotti, Sanford, & Levens, 1997; Michelotti, Tomonaga, Krutzsch, & Levens, 1995; Michelotti et al., 1996). The *HSP70I* gene is upregulated by an increase in temperature through the transcriptional activity of heat shock factor 1 (HSF1). The transcription factor HSF2 binds specifically the hsp70 promoter in mitosis and is responsible for the maintenance of an open chromatin conformation during cell division. HSF2 has two main mitotic interactors: CAP-G and protein phosphatase 2A (PP2A) (Xing et al., 2005). CAP-G is a subunit of the condensin complex (see above), and the complex has a function, which is the organization of mitotic chromosome structure as we have seen before. The activity of the complex is also regulated during cell cycle by phosphorylation. The condensin complex is phosphorylated by several mitotic kinases such as CDK1, Aurora B/Ipl1 (Giet & Glover, 2001; Hagstrom et al., 2002; Kaitna, Pasierbek, Jantsch, Loidl, & Glotzer, 2002; Lavoie, Hogan, & Koshland, 2004; Lipp, Hirota, Poser, & Peters, 2007; Ono, Fang, Spector, & Hirano, 2004; Petersen & Hagan, 2003; St-Pierre et al., 2009; Takemoto, Kimura, Yanagisawa, Yokoyama, & Hanaoka, 2006; Takemoto et al., 2007).

Some studies have suggested that a member of the PPP family of serine–threonine phosphatases is involved in dephosphorylating the condensin subunits (Takemoto et al., 2009; Xing et al., 2008; Yeong et al., 2003). Therefore, at the *hsp70* promoter during mitosis HSF2 bookmarks the *hsp70i gene* and locally recruits PP2A. PP2A in turn, dephosphorylates one or more subunits of the condensin complex, thus inactivating it and inhibiting local DNA supercoiling. This would generate a local less compact chromatin region even in mitosis. The biological importance of this is that loss of HSF2-mediated bookmarking reduces the ability of cells to induce hsp70i gene expression and to survive stress (Xing et al., 2008). Using a similar mechanism, the transcription factor TBP, which has also been shown to act as a mitotic bookmark, causes decompaction of its target gene by the local recruitment of PP2A to inactivate condensin action. Therefore, local inactivation of condensin appears to be one of the features adopted in gene bookmarking. We can envisage that by blocking a higher level of condensation in mitosis it could facilitate the reassembly of the transcription machinery at a specific region after cell division.

Could the generation of an open chromatin environment use other strategies rather than blocking local mitotic chromosome organization?

Not all signatures of active gene expression are erased during mitosis. For example, acetylation and methylation of histones are partially or completely maintained during mitosis (Aoto, Saitoh, Sakamoto, Watanabe, & Nakao,

2008; Blobel et al., 2009; Bonenfant et al., 2007; Kouskouti & Talianidis, 2005; Kruhlak et al., 2001; McManus, Biron, Heit, Underhill, & Hendzel, 2006; McManus & Hendzel, 2006; Sasaki, Ito, Nishino, Khochbin, & Yoshida, 2009; Valls, Sanchez-Molina, & Martinez–Balbas, 2005; Zaidi et al., 2003) and histone variants that mark active genes remain localized at the promoter regions (Bruce et al., 2005; Kelly, Miranda, et al., 2010). Therefore, the maintenance of a specific histone modification could also be used in principle as a mean for gene bookmarking. To address this question, Spector and colleagues have designed a system where they can induce specifically transcription of a locus and follow its silencing and reactivation in the transition to mitosis and to the next G1. In the specific experiment, Spector and colleagues found that the locus bookmarking was mediated by H4K5Ac (H4 lysine 5 acetylation). This particular modification persists in mitosis and is recognized by the bromodomain protein 4 (BRD4). Also in this case, the recruitment of BRD4 correlates with decompaction of chromatin at the locus.

Blobel and colleagues made a very important observation on the role of gene bookmarking in the maintenance of a normal hematopoietic program. In this system, a DNA-binding factor, GATA1 remains bound during mitosis to a subset of its target genes. The mitotic GATA1 preferentially occupies genes encoding lineage-specific transcription factors and the removal of GATA1 from mitotic chromatin, delays reactivation of these genes in daughter cells. In contrast to MLL, which preferentially marks highly expressed and housekeeping genes during mitosis (Blobel et al., 2009), GATA1 mitotic binding is not correlated with the level of transcription.

Therefore, the concept of gene bookmarking is clearly very important for cell division but the strategies that are used appear to be quite diverse in terms of signatures and effectors. This could just represent a sample of a plethora of different possibilities that are used according to the specific differentiation lineage or reactivation timing. More studies using different models should clarify which are the prevailing mechanisms and when are they preferentially adopted.

5. GOING BACK TO INTERPHASE: CHROMOSOME DECONDENSATION

We have dealt so far with changes that occur at the chromatin level during the transition from G2 to mitosis and during chromosome alignment; these reorganizations need to be reversed in the second part of cell division (mitotic exit) to give rise to the G1 interphase nucleus. During this part of mitosis, the segregated chromatids will reach the opposite spindle poles, the poles will move apart, the nuclear lamina will reform around the segregated genomes, the chromatin will decondense by leaving the rod-shaped structures typical of mitosis, and finally each chromosome will find its respective position within the interphase nucleus.

However, it is important that chromosome decondensation does not occur during the migration of the sister chromatids to the poles and until the nuclear lamina has reformed. Therefore, the maintenance of a compact chromosome structure is important for allowing the final stage of cell division (cytokinesis) to complete. In the past few years, we have gained information on how this is sustained and the responsible molecular machineries. During studies aimed to analyze the role of chromosome scaffold proteins in chromosome segregation, it was noted that the maximum compaction of the chromosomes during mitosis occurs in late mitosis, in anaphase (Vagnarelli et al., 2006). Although chromosomes lacking the condensin complex segregate aberrantly with trails of chromatin bridging between the two daughter cells, the chromatin hypercompaction typical of the anaphase chromosomes, is not lost. This observation was confirmed by using different imaging systems to measure chromatin compaction, and therefore, it has become an established characteristic of chromosome behavior in mitosis (Lleres et al., 2009; Mora–Bermudez, Gerlich, & Ellenberg, 2007).

In search for the factors that are responsible for the anaphase chromatid shortening, Ellenberg and colleagues identified that the localized Aurora B activity at the central spindle is necessary for anaphase chromatid shortening (Mora–Bermudez et al., 2007). These authors also noticed that when the shortening was perturbed by Aurora inhibition, abnormal multilobed G1 nuclei were formed instead of the smooth ellipsoid nuclei observed in control experiments. This observation suggests that the anaphase shortening could be an important mechanism that allows the proper organization of the interphase nucleus to be reestablished. However, details on interphase chromosome territories organization have not been analyzed so far. If Aurora B kinase activity is the mediator of this process, which are the substrates? Up-to-date these substrates are still mysterious. Another indication of the importance of anaphase chromatin hypercompaction came from studies in yeast. By manipulating chromosome length and structure, Neurohr and colleagues demonstrated that yeast cells adjust the compaction of chromosomes to secure their segregation by the spindle. Again they identified Aurora kinase as the effector of the signal (Neurohr et al., 2011).

Here a question arises: why chromosome compaction is still very important during these stages of cell division? If anaphase segregation is perturbed, for example, by having telomere dysfunctions, dicentric chromosomes, or incomplete chromosome decatenation, this can produce chromatin bridges between the segregated genomic masses. Anaphase segregation errors occur at low frequency in normal cells but at higher incidence in aging tissues and cancer cells. This represents a great danger toward chromosome instability since either the cleavage of the DNA trapped in the bridge or the abortion of cell division will have as end products an abnormal genomic content. Human cells containing chromosome bridges significantly delayed abscission by the assembly of stable actin-rich intercellular canals Therefore, it appears that completion of chromosome segregation and abscission timing are temporally coordinated and again this depends on sustained Aurora B activation. In this case, however, it appears to be mediated by the chromatin itself since just an amorphous physical barrier is not sufficient to trigger the response (Steigemann et al., 2009). It is therefore possible that Aurora B is also directly activated by binding to chromatin, and the observation that actually chromatin trapped in a bridge is still highly phosphorylated compared to the already segregated chromosome can support this hypothesis (Regnier et al., 2005; Vagnarelli et al., 2006) (Fig. 6.5: 10–12). The abscission checkpoint therefore provides an important safeguard mechanism against genomic instability and cancer.

How chromosome decondensation then occurs? It is quite well established that a requirement for mitotic exit is the degradation of B1-type cyclin and inactivation of CDK (Wolf, Wandke, Isenberg, & Geley, 2006). The expression of a nondegradable cyclin B1 causes mitotic arrest and blocks chromosome decondensation (Wheatley et al., 1997), and the over-expression Cyclin B1 produces an arrest in different stages of mitosis depending on the expression levels with the lowest levels being sufficient to block completion of cytokinesis and chromosome decondensation.

From studies in *Drosophila*, it appears that a sequential degradation *of* cyclins (A, B, and B3) is necessary for the timely progression of mitotic events; CyclinB3 appears to be the last to be degraded, and this degradation is necessary to chromosome condensation to occur (Parry & O'Farrell, 2001). A similar result was obtained in vertebrates where overexpression of CyclinB3 in a mutant for condensin not only rescued the anaphase chromosome bridging phenotypes typical of condensin-depleted cells but also prevented chromosome decondensation (Vagnarelli et al., 2006) (Fig. 6.5: 7–9). CDK inactivation is sufficient to trigger mitotic exit even in

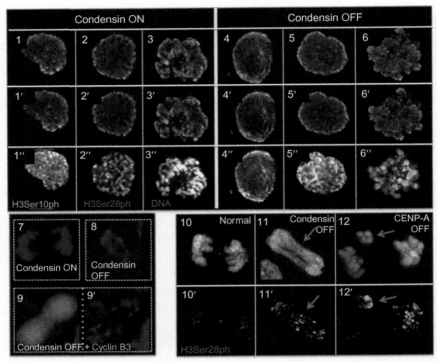

Figure 6.5 1–6: Stages of mitotic chromosome condensation and some histone modifications that are associated with the process. 1–3 normal cells: signs of chromosome condensation are already visible at the first appearance of phosphorylation on histone H3 at Ser28 (1–1″), progresses further with recognizable chromosome formation when phosphorylation on histone H3 at Ser28 is completed. In cells without condensin, there is a delay in the earliest stages of chromosome condensation with a clear sign of condensin structures only at the nuclear envelope breakdown (5–5′). 7–9: Cells lacking condensin present abnormal segregation in anaphase with the appearance of chromatin bridges that persist in late anaphase/telophase (8). This phenotype is alleviated by expressing high levels of CyclinB3 in anaphase (green in 9); compare the anaphase chromatin in 8 with the one in 9′. 10–12: Chromatin that does not segregate correctly in anaphase maintains a high level of phosphorylation on histone H3 (11, 11′; 12, 12′) and this could represent the signal for blocking cell division. See text for discussion. (See Color Insert.)

nocodazole-blocked cells and to induce chromosome decondensation and the reestablishment of the interphase status (Paulson, 2007).

Accumulating evidence indicates that, in vertebrates, protein phosphatases (PP1 and PP2A) are key players in mitotic exit (Asencio et al., 2012; Qian et al., 2011; Schmitz et al., 2010; Vagnarelli, 2012; Vagnarelli et al.,

2011; Wurzenberger et al., 2012). In this direction, several lines of evidence are converging to a pivotal role of PP1 in the modulation of chromatin decondensation and nuclear organization. PP1 catalytic subunit binds to targeting or regulatory subunits to both reach a substrate and achieve specificity. PP1 associates with mitotic chromatin in anaphase (Trinkle-Mulcahy et al., 2003), and the major chromatin-targeting subunit so far identified is Repo-Man (Trinkle-Mulcahy et al., 2006). Overexpression of a dominant-negative Repo-Man mutant that is capable to target to the anaphase chromosomes but does not bind PP1 (i.e., it displaces PP1 from the anaphase chromosomes) can again recues the anaphase bridges due to lack of condensin in a manner similar to the overexpression of Cyclin B3 (Vagnarelli et al., 2006). The targeting of Repo-Man in early anaphase is important for the dephosphorylation of histone H3 at Thr3, Ser10, and Ser28 (Vagnarelli et al., 2011; Qian et al., 2011; Wurzenberger et al., 2012). Although these modifications do not seem to be directly involved in mediating chromosome condensation, their removal is important to the accessibility of other proteins that directly or indirectly can mediate chromatin opening and promote decondensation.

Further studies have also identified that Repo-Man is not only involved in anaphase chromatin reorganization but also in nuclear envelope reformation. These observations suggest the hypothesis that these two late mitotic events can be coordinated (Vagnarelli et al., 2011; Vagnarelli & Earnshaw, 2012). A link between the two phenomena has also been observed in a knockdown study of the INner Membrane protein SUN1 (one of the earliest INM factors to associate with segregated daughter chromosomes in anaphase in human cells). Knockdown of hsSUN1 leads to hypoacetylated histones and delayed decondensation of chromosomes at the end of mitosis. This seems to be mediated by a HAT protein, hALP. In this experimental system, decondensation appears to be marked by several acetylated lysines in H2B and H4 including Lys15 of H2B and Lys8, Lys12, and Lys16 of H4 (Chi, Haller, Peloponese, & Jeang, 2007). hALP and hSUN1 can interact and this links seems to be brought about by the reformation of the nuclear envelope. If this explains a local chromatin remodeling/decondensation at the nuclear periphery, it does not explain how global chromatin decondensation is brought about.

Another phosphatase targeting subunit that has been suggested to be important for chromosome decondensation is PNUTS; it is exclusively nuclear in interphase and colocalizes with chromatin during telophase (Allen, Kwon, Nairn, & Greengard, 1998).

PNUTS is targeted to the reforming nucleus in telophase after reassembly of nuclear membranes and, in cell-free system, recombinant PNUTS promotes decondensation of purified mitotic chromosomes in a PP1-dependent manner. Since H3S10 dephosphorylation precedes targeting of PNUTS to reforming nuclei, it appears to be downstream of PP1 targeting and SUN1 pathway.

Although the emerging picture is that phosphatases are the keys to unravel mitotic chromatin, there are other observations that are a bit out or, maybe, we do not see a link with at the present moment. This relates to the phenotype observed in patients affected by the microcephaly syndrome. One of the genes found mutated in those patients is microcephalyn (MCPH1) and the remarkable phenotype observed in the cells from those patients is both prematurely condensed chromosomes, and highly delayed chromosome decondensation after cytokinesis is completed but with normal nuclear envelope reformation (Jackson et al., 1998, 2002). The molecular reason for the delayed decondensation is not known but in future years it will be possible to put this gene in the context of the chromosome decondensation pathways.

ACKNOWLEDGMENTS

I am really grateful to Dr. A. R. Mitchell for discussions and critical comments on the chapter. I would also like to thank, in particular, Prof. W. C. Earnshaw (University of Edinburgh, UK), Dr. S. Ohta (Kochi University, Japan), Dr. D. Booth (University of Edinburgh, UK), Dr. S. Tosi, and Prof. S. Saccone (University of Catania, Italy) for providing images for the figures.

REFERENCES

Adolph, K. W., Cheng, S. M., & Laemmli, U. K. (1977). Role of nonhistone proteins in metaphase chromosome structure. Cell, 12, 805–816.
Ainsztein, A. M., Kandels-Lewis, S. E., Mackay, A. M., & Earnshaw, W. C. (1998). INCENP centromere and spindle targeting: Identification of essential conserved motifs and involvement of heterochromatin protein HP1. The Journal of Cell Biology, 143, 1763–1774.
Akiyoshi, B., Sarangapani, K. K., Powers, A. F., Nelson, C. R., Reichow, S. L., Arellano-Santoyo, H., et al. (2010). Tension directly stabilizes reconstituted kinetochore-microtubule attachments. Nature, 468, 576–579.
Allen, P. B., Kwon, Y. G., Nairn, A. C., & Greengard, P. (1998). Isolation and characterization of PNUTS, a putative protein phosphatase 1 nuclear targeting subunit. The Journal of Biological Chemistry, 273, 4089–4095.
Alonso, A., Hasson, D., Cheung, F., & Warburton, P. E. (2010). A paucity of heterochromatin at functional human neocentromeres. Epigenetics and Chromatin, 3, 6.
Amano, M., Suzuki, A., Hori, T., Backer, C., Okawa, K., Cheeseman, I. M., et al. (2009). The CENP-S complex is essential for the stable assembly of outer kinetochore structure. The Journal of Cell Biology, 186, 173–182.

Aoto, T., Saitoh, N., Sakamoto, Y., Watanabe, S., & Nakao, M. (2008). Polycomb group protein-associated chromatin is reproduced in post-mitotic G1 phase and is required for S phase progression. *The Journal of Biological Chemistry*, *283*, 18905–18915.

Asencio, C., Davidson, I. F., Santarella-Mellwig, R., Ly-Hartig, T. B., Mall, M., Wallenfang, M. R., et al. (2012). Coordination of kinase and phosphatase activities by Lem4 enables nuclear envelope reassembly during mitosis. *Cell*, *150*, 122–135.

Bachant, J., Alcasabas, A., Blat, Y., Kleckner, N., & Elledge, S. J. (2002). The SUMO-1 iso-peptidase Smt4 is linked to centromeric cohesion through SUMO-1 modification of DNA topoisomerase II. *Molecular Cell*, *9*, 1169–1182.

Bai, H., Sun, M., Ghosh, P., Hatfull, G. F., Grindley, N. D., & Marko, J. F. (2011). Single-molecule analysis reveals the molecular bearing mechanism of DNA strand exchange by a serine recombinase. *Proceedings of the National Academy of Sciences of the United States of America*, *108*, 7419–7424.

Belmont, A. S., & Bruce, K. (1994). Visualization of G1 chromosomes: A folded, twisted, supercoiled chromonema model of interphase chromatid structure. *The Journal of Cell Biology*, *127*, 287–302.

Belmont, A. S., Sedat, J. W., & Agard, D. A. (1987). A three-dimensional approach to mitotic chromosome structure: Evidence for a complex hierarchical organization. *The Journal of Cell Biology*, *105*, 77–92.

Black, B. E., & Cleveland, D. W. (2011). Epigenetic centromere propagation and the nature of CENP-a nucleosomes. *Cell*, *144*, 471–479.

Blobel, G. A., Kadauke, S., Wang, E., Lau, A. W., Zuber, J., Chou, M. M., et al. (2009). A reconfigured pattern of MLL occupancy within mitotic chromatin promotes rapid tran-scriptional reactivation following mitotic exit. *Molecular Cell*, *36*, 970–983.

Bonenfant, D., Towbin, H., Coulot, M., Schindler, P., Mueller, D. R., & van Oostrum, J. (2007). Analysis of dynamic changes in post-translational modifications of human his-tones during cell cycle by mass spectrometry. *Molecular and Cellular Proteomics*, *6*, 1917–1932.

Boy de la Tour, E., & Laemmli, U. K. (1988). The metaphase scaffold is helically folded: Sister chromatids have predominantly opposite helical handedness. *Cell*, *55*, 937–944.

Bruce, K., Myers, F. A., Mantouvalou, E., Lefevre, P., Greaves, I., Bonifer, C., et al. (2005). The replacement histone H2A.Z in a hyperacetylated form is a feature of active genes in the chicken. *Nucleic Acids Research*, *33*, 5633–5639.

Bui, M., Dimitriadis, E. K., Hoischen, C., An, E., Quenet, D., Giebe, S., et al. (2012). Cell-cycle-dependent structural transitions in the human CENP-A nucleosome in vivo. *Cell*, *150*, 317–326.

Carmena, M., & Earnshaw, W. C. (2006). INCENP at the kinase crossroads. *Nature Cell Biology*, *8*, 110–111.

Carmena, M., Ruchaud, S., & Earnshaw, W. C. (2009). Making the Auroras glow: Regu-lation of Aurora A and B kinase function by interacting proteins. *Current Opinion in Cell Biology*, *21*, 796–805.

Chen, D., Hinkley, C. S., Henry, R. W., & Huang, S. (2002). TBP dynamics in living human cells: Constitutive association of TBP with mitotic chromosomes. *Molecular Biol-ogy of the Cell*, *13*, 276–284.

Chi, Y. H., Haller, K., Peloponese, J. M., Jr., & Jeang, K. T. (2007). Histone acetyltransferase hALP and nuclear membrane protein hsSUN1 function in de-condensation of mitotic chromosomes. *The Journal of Biological Chemistry*, *282*, 27447–27458.

Christova, R., & Oelgeschlager, T. (2002). Association of human TFIID-promoter com-plexes with silenced mitotic chromatin in vivo. *Nature Cell Biology*, *4*, 79–82.

Daban, J. R., & Bermudez, A. (1998). Interdigitated solenoid model for compact chromatin fibers. *Biochemistry*, *37*, 4299–4304.

Dai, J., Sullivan, B. A., & Higgins, J. M. (2006). Regulation of mitotic chromosome cohesion by Haspin and Aurora B. *Developmental Cell*, *11*, 741–750.

Dalal, Y., Furuyama, T., Vermaak, D., & Henikoff, S. (2007). Structure, dynamics, and evolution of centromeric nucleosomes. *Proceedings of the National Academy of Sciences of the United States of America*, *104*, 15974–15981.

Dalal, Y., Wang, H., Lindsay, S., & Henikoff, S. (2007). Tetrameric structure of centromeric nucleosomes in interphase Drosophila cells. *PLoS Biology*, *5*, e218.

Dekker, J. (2008). Mapping in vivo chromatin interactions in yeast suggests an extended chromatin fiber with regional variation in compaction. *The Journal of Biological Chemistry*, *283*, 34532–34540.

Dewar, H., Tanaka, K., Nasmyth, K., & Tanaka, T. U. (2004). Tension between two kinetochores suffices for their bi-orientation on the mitotic spindle. *Nature*, *428*, 93–97.

Dey, A., Ellenberg, J., Farina, A., Coleman, A. E., Maruyama, T., Sciortino, S., et al. (2000). A bromodomain protein, MCAP, associates with mitotic chromosomes and affects G(2)-to-M transition. *Molecular and Cellular Biology*, *20*, 6537–6549.

Drew, H. R., & Travers, A. A. (1985a). DNA bending and its relation to nucleosome positioning. *Journal of Molecular Biology*, *186*, 773–790.

Drew, H. R., & Travers, A. A. (1985b). Structural junctions in DNA: The influence of flanking sequence on nuclease digestion specificities. *Nucleic Acids Research*, *13*, 4445–4467.

Duncan, R., Bazar, L., Michelotti, G., Tomonaga, T., Krutzsch, H., Avigan, M., et al. (1994). A sequence-specific, single-strand binding protein activates the far upstream element of c-myc and defines a new DNA-binding motif. *Genes and Development*, *8*, 465–480.

Earnshaw, W. C., Halligan, B., Cooke, C. A., Heck, M. M., & Liu, L. F. (1985). Topoisomerase II is a structural component of mitotic chromosome scaffolds. *The Journal of Cell Biology*, *100*, 1706–1715.

Earnshaw, W. C., & Heck, M. M. (1985). Localization of topoisomerase II in mitotic chromosomes. *The Journal of Cell Biology*, *100*, 1716–1725.

Earnshaw, W. C., & Laemmli, U. K. (1983). Architecture of metaphase chromosomes and chromosome scaffolds. *The Journal of Cell Biology*, *96*, 84–93.

Egli, D., Birkhoff, G., & Eggan, K. (2008). Mediators of reprogramming: Transcription factors and transitions through mitosis. *Nature Reviews Molecular Cell Biology*, *9*, 505–516.

Finch, J. T., & Klug, A. (1976). Solenoidal model for superstructure in chromatin. *Proceedings of the National Academy of Sciences of the United States of America*, *73*, 1897–1901.

Finch, J. T., Lutter, L. C., Rhodes, D., Brown, R. S., Rushton, B., Levitt, M., et al. (1977). Structure of nucleosome core particles of chromatin. *Nature*, *269*, 29–36.

Fukagawa, T., Nogami, M., Yoshikawa, M., Ikeno, M., Okazaki, T., Takami, Y., et al. (2004). Dicer is essential for formation of the heterochromatin structure in vertebrate cells. *Nature Cell Biology*, *6*, 784–791.

Fussner, E., Ching, R. W., & Bazett-Jones, D. P. (2011). Living without 30nm chromatin fibers. *Trends in Biochemical Sciences*, *36*, 1–6.

Fussner, E., Djuric, U., Strauss, M., Hotta, A., Perez-Iratxeta, C., Lanner, F., et al. (2011). Constitutive heterochromatin reorganization during somatic cell reprogramming. *EMBO Journal*, *30*, 1778–1789.

Fussner, E., Strauss, M., Djuric, U., Li, R., Ahmed, K., Hart, M., et al. (2012). Open and closed domains in the mouse genome are configured as 10-nm chromatin fibres. *EMBO Reports 13*, 992–996.

Gascoigne, K. E., Takeuchi, K., Suzuki, A., Hori, T., Fukagawa, T., & Cheeseman, I. M. (2011). Induced ectopic kinetochore assembly bypasses the requirement for CENP-A nucleosomes. *Cell*, *145*, 410–422.

Gasser, S. M., Laroche, T., Falquet, J., Boy de la Tour, E., & Laemmli, U. K. (1986). Metaphase chromosome structure. Involvement of topoisomerase II. *Journal of Molecular Biology, 188*, 613–629.

Gavet, O., & Pines, J. (2010a). Activation of cyclin B1-Cdk1 synchronizes events in the nucleus and the cytoplasm at mitosis. *The Journal of Cell Biology, 189*, 247–259.

Gavet, O., & Pines, J. (2010b). Progressive activation of CyclinB1-Cdk1 coordinates entry to mitosis. *Developmental Cell, 18*, 533–543.

Gerlich, D., Hirota, T., Koch, B., Peters, J. M., & Ellenberg, J. (2006). Condensin I stabilizes chromosomes mechanically through a dynamic interaction in live cells. *Current Biology, 16*, 333–344.

Giet, R., & Glover, D. M. (2001). Drosophila aurora B kinase is required for histone H3 phosphorylation and condensin recruitment during chromosome condensation and to organize the central spindle during cytokinesis. *The Journal of Cell Biology, 152*, 669–682.

Goshima, G., Kiyomitsu, T., Yoda, K., & Yanagida, M. (2003). Human centromere chromatin protein hMis12, essential for equal segregation, is independent of CENP-A loading pathway. *The Journal of Cell Biology, 160*, 25–39.

Grewal, S. I., & Jia, S. (2007). Heterochromatin revisited. *Nature Reviews. Genetics, 8*, 35–46.

Groudine, M., & Weintraub, H. (1982). Propagation of globin DNAase I-hypersensitive sites in absence of factors required for induction: A possible mechanism for determination. *Cell, 30*, 131–139.

Guenatri, M., Bailly, D., Maison, C., & Almouzni, G. (2004). Mouse centric and pericentric satellite repeats form distinct functional heterochromatin. *The Journal of Cell Biology, 166*, 493–505.

Hagstrom, K. A., Holmes, V. F., Cozzarelli, N. R., & Meyer, B. J. (2002). *C. elegans* condensin promotes mitotic chromosome architecture, centromere organization, and sister chromatid segregation during mitosis and meiosis. *Genes and Development, 16*, 729–742.

Hansen, J. C. (2002). Conformational dynamics of the chromatin fiber in solution: Determinants, mechanisms, and functions. *Annual Review of Biophysics and Biomolecular Structure, 31*, 361–392.

Hauf, S., Roitinger, E., Koch, B., Dittrich, C. M., Mechtler, K., & Peters, J. M. (2005). Dissociation of cohesin from chromosome arms and loss of arm cohesion during early mitosis depends on phosphorylation of SA2. *PLoS Biology, 3*, e69.

Hauf, S., Waizenegger, I. C., & Peters, J. M. (2001). Cohesin cleavage by separase required for anaphase and cytokinesis in human cells. *Science, 293*, 1320–1323.

Higgins, J. M. (2010). Haspin: A newly discovered regulator of mitotic chromosome behavior. *Chromosoma, 119*, 137–147.

Hirano, T., Kobayashi, R., & Hirano, M. (1997). Condensins, chromosome condensation protein complexes containing XCAP-C, XCAP-E and a Xenopus homolog of the Drosophila Barren protein. *Cell, 89*, 511–521.

Hirota, T., Lipp, J. J., Toh, B. H., & Peters, J. M. (2005). Histone H3 serine 10 phosphorylation by Aurora B causes HP1 dissociation from heterochromatin. *Nature, 438*, 1176–1180.

Hori, T., Amano, M., Suzuki, A., Backer, C. B., Welburn, J. P., Dong, Y., et al. (2008). CCAN makes multiple contacts with centromeric DNA to provide distinct pathways to the outer kinetochore. *Cell, 135*, 1039–1052.

Hori, T., & Fukagawa, T. (2012). Establishment of the vertebrate kinetochores. *Chromosome Research, 20*, 547–561.

Hori, T., Okada, M., Maenaka, K., & Fukagawa, T. (2008). CENP-O class proteins form a stable complex and are required for proper kinetochore function. *Molecular Biology of the Cell, 19*, 843–854.

Howman, E. V., Fowler, K. J., Newson, A. J., Redward, S., MacDonald, A. C., Kalitsis, P., et al. (2000). Early disruption of centromeric chromatin organization in centromere protein A (Cenpa) null mice. *Proceedings of the National Academy of Sciences of the United States of America, 97,* 1148–1153.

Huang, H., Feng, J., Famulski, J., Rattner, J. B., Liu, S. T., Kao, G. D., et al. (2007). Tripin/hSgo2 recruits MCAK to the inner centromere to correct defective kinetochore attachments. *The Journal of Cell Biology, 177,* 413–424.

Hudson, D. F., Vagnarelli, P., Gassmann, R., & Earnshaw, W. C. (2003). Condensin is required for nonhistone protein assembly and structural integrity of vertebrate mitotic chromosomes. *Developmental Cell, 5,* 323–336.

Hughes, A. L., Jin, Y., Rando, O. J., & Struhl, K. (2012). A functional evolutionary approach to identify determinants of nucleosome positioning: A unifying model for establishing the genome-wide pattern. *Molecular Cell, 48,* 5–15.

Jackson, A. P., Eastwood, H., Bell, S. M., Adu, J., Toomes, C., Carr, I. M., et al. (2002). Identification of microcephalin, a protein implicated in determining the size of the human brain. *The American Journal of Human Genetics, 71,* 136–142.

Jackson, A. P., McHale, D. P., Campbell, D. A., Jafri, H., Rashid, Y., Mannan, J., et al. (1998). Primary autosomal recessive microcephaly (MCPH1) maps to chromosome 8p22-pter. *The American Journal of Human Genetics, 63,* 541–546.

Jaqaman, K., King, E. M., Amaro, A. C., Winter, J. R., Dorn, J. F., Elliott, H. L., et al. (2010). Kinetochore alignment within the metaphase plate is regulated by centromere stiffness and microtubule depolymerases. *The Journal of Cell Biology, 188,* 665–679.

Jeyaprakash, A. A., Klein, U. R., Lindner, D., Ebert, J., Nigg, E. A., & Conti, E. (2007). Structure of a Survivin-Borealin-INCENP core complex reveals how chromosomal passengers travel together. *Cell, 131,* 271–285.

Jiang, C., & Pugh, B. F. (2009a). A compiled and systematic reference map of nucleosome positions across the Saccharomyces cerevisiae genome. *Genome Biology, 10,* R109.

Jiang, C., & Pugh, B. F. (2009b). Nucleosome positioning and gene regulation: Advances through genomics. *Nature Reviews Genetics, 10,* 161–172.

Jin, W., Lamb, J. C., Zhang, W., Kolano, B., Birchler, J. A., & Jiang, J. (2008). Histone modifications associated with both A and B chromosomes of maize. *Chromosome Research, 16,* 1203–1214.

Kaitna, S., Pasierbek, P., Jantsch, M., Loidl, J., & Glotzer, M. (2002). The aurora B kinase AIR-2 regulates kinetochores during mitosis and is required for separation of homologous Chromosomes during meiosis. *Current Biology, 12,* 798–812.

Kanno, T., Kanno, Y., Siegel, R. M., Jang, M. K., Lenardo, M. J., & Ozato, K. (2004). Selective recognition of acetylated histones by bromodomain proteins visualized in living cells. *Molecular Cell, 13,* 33–43.

Kelly, A. E., Ghenoiu, C., Xue, J. Z., Zierhut, C., Kimura, H., & Funabiki, H. (2010). Survivin reads phosphorylated histone H3 threonine 3 to activate the mitotic kinase Aurora B. *Science, 330,* 235–239.

Kelly, T. K., Miranda, T. B., Liang, G., Berman, B. P., Lin, J. C., Tanay, A., et al. (2010). H2A.Z maintenance during mitosis reveals nucleosome shifting on mitotically silenced genes. *Molecular Cell, 39,* 901–911.

Kimura, K., Hirano, M., Kobayashi, R., & Hirano, T. (1998). Phosphorylation and activation of 13S condensin by Cdc2 in vitro. *Science, 282,* 487–490.

Kireeva, N., Lakonishok, M., Kireev, I., Hirano, T., & Belmont, A. S. (2004). Visualization of early chromosome condensation: A hierarchical folding, axial glue model of chromosome structure. *The Journal of Cell Biology, 166,* 775–785.

Kitajima, T. S., Kawashima, S. A., & Watanabe, Y. (2004). The conserved kinetochore protein shugoshin protects centromeric cohesion during meiosis. *Nature, 427,* 510–517.

Kiyomitsu, T., Iwasaki, O., Obuse, C., & Yanagida, M. (2010). Inner centromere formation requires hMis14, a trident kinetochore protein that specifically recruits HP1 to human chromosomes. *The Journal of Cell Biology*, *188*, 791–807.

Koch, B., Kueng, S., Ruckenbauer, C., Wendt, K. S., & Peters, J. M. (2008). The Suv39h-HP1 histone methylation pathway is dispensable for enrichment and protection of cohesin at centromeres in mammalian cells. *Chromosoma*, *117*, 199–210.

Komura, J., Ikehata, H., & Ono, T. (2007). Chromatin fine structure of the c-MYC insulator element/DNase I-hypersensitive site I is not preserved during mitosis. *Proceedings of the National Academy of Sciences of the United States of America*, *104*, 15741–15746.

Kornberg, R. D., & Lorch, Y. (1999a). Chromatin-modifying and -remodeling complexes. *Current Opinion in Genetics and Development*, *9*, 148–151.

Kornberg, R. D., & Lorch, Y. (1999b). Twenty-five years of the nucleosome, fundamental particle of the eukaryote chromosome. *Cell*, *98*, 285–294.

Kornberg, R. D., & Thomas, J. O. (1974). Chromatin structure; oligomers of the histones. *Science*, *184*, 865–868.

Kouskouti, A., & Talianidis, I. (2005). Histone modifications defining active genes persist after transcriptional and mitotic inactivation. *EMBO Journal*, *24*, 347–357.

Kruhlak, M. J., Hendzel, M. J., Fischle, W., Bertos, N. R., Hameed, S., Yang, X. J., et al. (2001). Regulation of global acetylation in mitosis through loss of histone acetyltransferases and deacetylases from chromatin. *The Journal of Biological Chemistry*, *276*, 38307–38319.

Kruithof, M., Chien, F. T., Routh, A., Logie, C., Rhodes, D., & van Noort, J. (2009). Single-molecule force spectroscopy reveals a highly compliant helical folding for the 30-nm chromatin fiber. *Nature Structural and Molecular Biology*, *16*, 534–540.

Kunitoku, N., Sasayama, T., Marumoto, T., Zhang, D., Honda, S., Kobayashi, O., et al. (2003). CENP-A phosphorylation by Aurora-A in prophase is required for enrichment of Aurora-B at inner centromeres and for kinetochore function. *Developmental Cell*, *5*, 853–864.

Lachner, M., O'Carroll, D., Rea, S., Mechtler, K., & Jenuwein, T. (2001). Methylation of histone H3 lysine 9 creates a binding site for HP1 proteins. *Nature*, *410*, 116–120.

Laemmli, U. K. (1978). Levels of organization of the DNA in eukaryotic chromosomes. *Pharmacological Reviews*, *30*, 469–476.

Laemmli, U. K., Cheng, S. M., Adolph, K. W., Paulson, J. R., Brown, J. A., & Baumbach, W. R. (1978). Metaphase chromosome structure: The role of nonhistone proteins. *Cold Spring Harbor Symposia on Quantitative Biology*, *42*(Pt. 1), 351–360.

Lampson, M. A., & Cheeseman, I. M. (2011). Sensing centromere tension: Aurora B and the regulation of kinetochore function. *Trends in Cell Biology*, *21*, 133–140.

Lavoie, B. D., Hogan, E., & Koshland, D. (2004). In vivo requirements for rDNA chromosome condensation reveal two cell-cycle-regulated pathways for mitotic chromosome folding. *Genes and Development*, *18*, 76–87.

Levens, D., Duncan, R. C., Tomonaga, T., Michelotti, G. A., Collins, I., Davis-Smyth, T., et al. (1997). DNA conformation, topology, and the regulation of c-myc expression. *Current Topics in Microbiology and Immunology*, *224*, 33–46.

Lewis, C. D., & Laemmli, U. K. (1982). Higher order metaphase chromosome structure: Evidence for metalloprotein interactions. *Cell*, *29*, 171–181.

Lieberman-Aiden, E., van Berkum, N. L., Williams, L., Imakaev, M., Ragoczy, T., Telling, A., et al. (2009). Comprehensive mapping of long-range interactions reveals folding principles of the human genome. *Science*, *326*, 289–293.

Lipp, J. J., Hirota, T., Poser, I., & Peters, J. M. (2007). Aurora B controls the association of condensin I but not condensin II with mitotic chromosomes. *Journal of Cell Science*, *120*, 1245–1255.

Liu, S., Tao, Y., Chen, X., & Cao, Y. (2012). The dynamic interplay in chromatin remodeling factors polycomb and trithorax proteins in response to DNA damage. *Molecular Biology Reports*, *39*, 6179–6185.

Liu, D., Vader, G., Vromans, M. J., Lampson, M. A., & Lens, S. M. (2009). Sensing chromosome bi-orientation by spatial separation of aurora B kinase from kinetochore substrates. *Science*, *323*, 1350–1353.

Lleres, D., James, J., Swift, S., Norman, D. G., & Lamond, A. I. (2009). Quantitative analysis of chromatin compaction in living cells using FLIM-FRET. *The Journal of Cell Biology*, *187*, 481–496.

Luger, K., & Richmond, T. J. (1998a). DNA binding within the nucleosome core. *Current Opinion in Structural Biology*, *8*, 33–40.

Luger, K., & Richmond, T. J. (1998b). The histone tails of the nucleosome. *Current Opinion in Genetics and Development*, *8*, 140–146.

Maeshima, K., Hihara, S., & Takata, H. (2010). New insight into the mitotic chromosome structure: Irregular folding of nucleosome fibers without 30-nm chromatin structure. *Cold Spring Harbor Symposia on Quantitative Biology*, *75*, 439–444.

Maeshima, K., & Laemmli, U. K. (2003). A two-step scaffolding model for mitotic chromosome assembly. *Developmental Cell*, *4*, 467–480.

Marsden, M. P., & Laemmli, U. K. (1979). Metaphase chromosome structure: Evidence for a radial loop model. *Cell*, *17*, 849–858.

Marshall, O. J., Chueh, A. C., Wong, L. H., & Choo, K. H. (2008). Neocentromeres: New insights into centromere structure, disease development, and karyotype evolution. *The American Journal of Human Genetics*, *82*, 261–282.

Martin, R. M., & Cardoso, M. C. (2010). Chromatin condensation modulates access and binding of nuclear proteins. *Federation of American Societies for Experimental Biology Journal*, *24*, 1066–1072.

Martinez-Balbas, M. A., Dey, A., Rabindran, S. K., Ozato, K., & Wu, C. (1995). Displacement of sequence-specific transcription factors from mitotic chromatin. *Cell*, *83*, 29–38.

Mazumdar, M., Sundareshan, S., & Misteli, T. (2004). Human chromokinesin KIF4A functions in chromosome condensation and segregation. *The Journal of Cell Biology*, *166*, 613–620.

McManus, K. J., Biron, V. L., Heit, R., Underhill, D. A., & Hendzel, M. J. (2006). Dynamic changes in histone H3 lysine 9 methylations: Identification of a mitosis-specific function for dynamic methylation in chromosome congression and segregation. *The Journal of Biological Chemistry*, *281*, 8888–8897.

McManus, K. J., & Hendzel, M. J. (2006). The relationship between histone H3 phosphorylation and acetylation throughout the mammalian cell cycle. *Biochemistry and Cell Biology*, *84*, 640–657.

Michelotti, G. A., Michelotti, E. F., Pullner, A., Duncan, R. C., Eick, D., & Levens, D. (1996). Multiple single-stranded cis elements are associated with activated chromatin of the human c-myc gene in vivo. *Molecular and Cellular Biology*, *16*, 2656–2669.

Michelotti, E. F., Sanford, S., & Levens, D. (1997). Marking of active genes on mitotic chromosomes. *Nature*, *388*, 895–899.

Michelotti, E. F., Tomonaga, T., Krutzsch, H., & Levens, D. (1995). Cellular nucleic acid binding protein regulates the CT element of the human c-myc protooncogene. *The Journal of Biological Chemistry*, *270*, 9494–9499.

Mirkovitch, J., Mirault, M. E., & Laemmli, U. K. (1984). Organization of the higher-order chromatin loop: Specific DNA attachment sites on nuclear scaffold. *Cell*, *39*, 223–232.

Mora-Bermudez, F., Gerlich, D., & Ellenberg, J. (2007). Maximal chromosome compaction occurs by axial shortening in anaphase and depends on Aurora kinase. *Nature Cell Biology*, *9*, 822–831.

Nakano, M., Cardinale, S., Noskov, V. N., Gassmann, R., Vagnarelli, P., Kandels-Lewis, S., et al. (2008). Inactivation of a human kinetochore by specific targeting of chromatin modifiers. *Developmental Cell*, *14*, 507–522.

Neurohr, G., Naegeli, A., Titos, I., Theler, D., Greber, B., Diez, J., et al. (2011). A midzone-based ruler adjusts chromosome compaction to anaphase spindle length. *Science*, *332*, 465–468.

Nicklas, R. B., & Koch, C. A. (1969). Chromosome micromanipulation. 3. Spindle fiber tension and the reorientation of mal-oriented chromosomes. *The Journal of Cell Biology*, *43*, 40–50.

Nielsen, P. R., Nietlispach, D., Mott, H. R., Callaghan, J., Bannister, A., Kouzarides, T., et al. (2002). Structure of the HP1 chromodomain bound to histone H3 methylated at lysine 9. *Nature*, *416*, 103–107.

Nishino, Y., Eltsov, M., Joti, Y., Ito, K., Takata, H., Takahashi, Y., et al. (2012). Human mitotic chromosomes consist predominantly of irregularly folded nucleosome fibres without a 30-nm chromatin structure. *EMBO Journal*, *31*, 1644–1653.

Oegema, K., Desai, A., Rybina, S., Kirkham, M., & Hyman, A. A. (2001). Functional analysis of kinetochore assembly in Caenorhabditis elegans. *The Journal of Cell Biology*, *153*, 1209–1226.

Ohta, S., Bukowski-Wills, J. C., Sanchez-Pulido, L., Alves Fde, L., Wood, L., Chen, Z. A., et al. (2010). The protein composition of mitotic chromosomes determined using multiclassifier combinatorial proteomics. *Cell*, *142*, 810–821.

Oliveira, R. A., Heidmann, S., & Sunkel, C. E. (2007). Condensin I binds chromatin early in prophase and displays a highly dynamic association with Drosophila mitotic chromosomes. *Chromosoma*, *116*, 259–274.

Ono, T., Fang, Y., Spector, D. L., & Hirano, T. (2004). Spatial and temporal regulation of Condensins I and II in mitotic chromosome assembly in human cells. *Molecular Biology of the Cell*, *15*, 3296–3308.

Ono, T., Losada, A., Hirano, M., Myers, M. P., Neuwald, A. F., & Hirano, T. (2003). Differential contributions of condensin I and condensin II to mitotic chromosome architecture in vertebrate cells. *Cell*, *115*, 109–121.

Orth, M., Mayer, B., Rehm, K., Rothweiler, U., Heidmann, D., Holak, T. A., et al. (2011). Shugoshin is a Mad1/Cdc20-like interactor of Mad2. *EMBO Journal*, *30*, 2868–2880.

Parry, D. H., & O'Farrell, P. H. (2001). The schedule of destruction of three mitotic cyclins can dictate the timing of events during exit from mitosis. *Current Biology*, *11*, 671–683.

Paulson, J. R. (2007). Inactivation of Cdk1/Cyclin B in metaphase-arrested mouse FT210 cells induces exit from mitosis without chromosome segregation or cytokinesis and allows passage through another cell cycle. *Chromosoma*, *116*, 215–225.

Paulson, J. R., & Laemmli, U. K. (1977). The structure of histone-depleted metaphase chromosomes. *Cell*, *12*, 817–828.

Petersen, J., & Hagan, I. M. (2003). S. pombe aurora kinase/survivin is required for chromosome condensation and the spindle checkpoint attachment response. *Current Biology*, *13*, 590–597.

Poirier, M., Eroglu, S., Chatenay, D., & Marko, J. F. (2000). Reversible and irreversible unfolding of mitotic newt chromosomes by applied force. *Molecular Biology of the Cell*, *11*, 269–276.

Poirier, M. G., Eroglu, S., & Marko, J. F. (2002). The bending rigidity of mitotic chromosomes. *Molecular Biology of the Cell*, *13*, 2170–2179.

Poirier, M. G., & Marko, J. F. (2002a). Micromechanical studies of mitotic chromosomes. *Journal of Muscle Research and Cell Motility*, *23*, 409–431.

Poirier, M. G., & Marko, J. F. (2002b). Mitotic chromosomes are chromatin networks without a mechanically contiguous protein scaffold. *Proceedings of the National Academy of Sciences of the United States of America*, *99*, 15393–15397.

Qian, J., Lesage, B., Beullens, M., Van Eynde, A., & Bollen, M. (2011). PP1/Repo-man dephosphorylates mitotic histone H3 at T3 and regulates chromosomal aurora B targeting. *Current Biology, 21*, 766–773.

Radman-Livaja, M., & Rando, O. J. (2010). Nucleosome positioning: How is it established, and why does it matter? *Developmental Biology, 339*, 258–266.

Rattner, J. B., Hendzel, M. J., Furbee, C. S., Muller, M. T., & Bazett-Jones, D. P. (1996). Topoisomerase II alpha is associated with the mammalian centromere in a cell cycle- and species-specific manner and is required for proper centromere/kinetochore structure. *The Journal of Cell Biology, 134*, 1097–1107.

Rattner, J. B., & Lin, C. C. (1985). Radial loops and helical coils coexist in metaphase chromosomes. *Cell, 42*, 291–296.

Razin, S. V. (1996). Functional architecture of chromosomal DNA domains. *Critical Reviews in Eukaryotic Gene Expression, 6*, 247–269.

Regnier, V., Vagnarelli, P., Fukagawa, T., Zerjal, T., Burns, E., Trouche, D., et al. (2005). CENP-A is required for accurate chromosome segregation and sustained kinetochore association of BubR1. *Molecular and Cellular Biology, 25*, 3967–3981.

Ribeiro, S. A., Gatlin, J. C., Dong, Y., Joglekar, A., Cameron, L., Hudson, D. F., et al. (2009). Condensin regulates the stiffness of vertebrate centromeres. *Molecular Biology of the Cell, 20*, 2371–2380.

Ribeiro, S. A., Vagnarelli, P., Dong, Y., Hori, T., McEwen, B. F., Fukagawa, T., et al. (2010). A super-resolution map of the vertebrate kinetochore. *Proceedings of the National Academy of Sciences of the United States of America, 107*, 10484–10489.

Robinson, P. J., Fairall, L., Huynh, V. A., & Rhodes, D. (2006). EM measurements define the dimensions of the "30-nm" chromatin fiber: Evidence for a compact, interdigitated structure. *Proceedings of the National Academy of Sciences of the United States of America, 103*, 6506–6511.

Robinson, P. J., & Rhodes, D. (2006). Structure of the '30 nm' chromatin fibre: A key role for the linker histone. *Current Opinion in Structural Biology, 16*, 336–343.

Ruchaud, S., Carmena, M., & Earnshaw, W. C. (2007). Chromosomal passengers: Conducting cell division. *Nature Reviews Molecular Cell Biology, 8*, 798–812.

Saitoh, Y., & Laemmli, U. K. (1994). Metaphase chromosome structure: Bands arise from a differential folding path of the highly AT-rich scaffold. *Cell, 76*, 609–622.

Samejima, K., Samejima, I., Vagnarelli, P., Ogawa, H., Vargiu, G., Kelly, D. A., et al. (2012). Mitotic chromosomes are compacted laterally by KIF4 and Condensin and axially by Topoisomerase IIa. *The Journal of Cell Biology, 199*, 755–770.

Samoshkin, A., Arnaoutov, A., Jansen, L. E., Ouspenski, I., Dye, L., Karpova, T., et al. (2009). Human condensin function is essential for centromeric chromatin assembly and proper sister kinetochore orientation. *PLoS One, 4*, e6831.

Santaguida, S., & Musacchio, A. (2009). The life and miracles of kinetochores. *EMBO Journal, 28*, 2511–2531.

Sarge, K. D., & Park-Sarge, O. K. (2009). Mitotic bookmarking of formerly active genes: Keeping epigenetic memories from fading. *Cell Cycle, 8*, 818–823.

Sasaki, K., Ito, T., Nishino, N., Khochbin, S., & Yoshida, M. (2009). Real-time imaging of histone H4 hyperacetylation in living cells. *Proceedings of the National Academy of Sciences of the United States of America, 106*, 16257–16262.

Schmitz, M. H., Held, M., Janssens, V., Hutchins, J. R., Hudecz, O., Ivanova, E., et al. (2010). Live-cell imaging RNAi screen identifies PP2A-B55alpha and importin-beta1 as key mitotic exit regulators in human cells. *Nature Cell Biology, 12*, 886–893.

Sedat, J., & Manuelidis, L. (1978). A direct approach to the structure of eukaryotic chromosomes. *Cold Spring Harbor Symposia on Quantitative Biology, 42*(Pt. 1), 331–350.

Segil, N., Guermah, M., Hoffmann, A., Roeder, R. G., & Heintz, N. (1996). Mitotic regulation of TFIID: Inhibition of activator-dependent transcription and changes in subcellular localization. *Genes and Development, 10*, 2389–2400.

Sessa, F., Mapelli, M., Ciferri, C., Tarricone, C., Areces, L. B., Schneider, T. R., et al. (2005). Mechanism of Aurora B activation by INCENP and inhibition by hesperadin. *Molecular Cell, 18*, 379–391.

Shi, J., & Dawe, R. K. (2006). Partitioning of the maize epigenome by the number of methyl groups on histone H3 lysines 9 and 27. *Genetics, 173*, 1571–1583.

Shivaraju, M., Unruh, J. R., Slaughter, B. D., Mattingly, M., Berman, J., & Gerton, J. L. (2012). Cell-cycle-coupled structural oscillation of centromeric nucleosomes in yeast. *Cell, 150*, 304–316.

Spence, J. M., Phua, H. H., Mills, W., Carpenter, A. J., Porter, A. C., & Farr, C. J. (2007). Depletion of topoisomerase IIalpha leads to shortening of the metaphase interkinetochore distance and abnormal persistence of PICH-coated anaphase threads. *Journal of Cell Science, 120*, 3952–3964.

Steigemann, P., Wurzenberger, C., Schmitz, M. H., Held, M., Guizetti, J., Maar, S., et al. (2009). Aurora B-mediated abscission checkpoint protects against tetraploidization. *Cell, 136*, 473–484.

Stephens, A. D., Haase, J., Vicci, L., Taylor, R. M., 2nd., & Bloom, K. (2011). Cohesin, condensin, and the intramolecular centromere loop together generate the mitotic chromatin spring. *The Journal of Cell Biology, 193*, 1167–1180.

St-Pierre, J., Douziech, M., Bazile, F., Pascariu, M., Bonneil, E., Sauve, V., et al. (2009). Polo kinase regulates mitotic chromosome condensation by hyperactivation of condensin DNA supercoiling activity. *Molecular Cell, 34*, 416–426.

Strick, R., Strissel, P. L., Gavrilov, K., & Levi-Setti, R. (2001). Cation-chromatin binding as shown by ion microscopy is essential for the structural integrity of chromosomes. *The Journal of Cell Biology, 155*, 899–910.

Stumpff, J., Wagenbach, M., Franck, A., Asbury, C. L., & Wordeman, L. (2012). Kif18A and chromokinesins confine centromere movements via microtubule growth suppression and spatial control of kinetochore tension. *Developmental Cell, 22*, 1017–1029.

Sugimoto, K., Tasaka, H., & Dotsu, M. (2001). Molecular behavior in living mitotic cells of human centromere heterochromatin protein HPLalpha ectopically expressed as a fusion to red fluorescent protein. *Cell Structure and Function, 26*, 705–718.

Sullivan, B. A., & Karpen, G. H. (2004). Centromeric chromatin exhibits a histone modification pattern that is distinct from both euchromatin and heterochromatin. *Nature Structural and Molecular Biology, 11*, 1076–1083.

Sumara, I., Vorlaufer, E., Stukenberg, P. T., Kelm, O., Redemann, N., Nigg, E. A., et al. (2002). The dissociation of cohesin from chromosomes in prophase is regulated by Polo-like kinase. *Molecular Cell, 9*, 515–525.

Sun, M., Kawamura, R., & Marko, J. F. (2011). Micromechanics of human mitotic chromosomes. *Physical Biology, 8*, 015003.

Swedlow, J. R., & Hirano, T. (2003). The making of the mitotic chromosome: Modern insights into classical questions. *Molecular Cell, 11*, 557–569.

Szerlong, H. J., Prenni, J. E., Nyborg, J. K., & Hansen, J. C. (2010). Activator-dependent p300 acetylation of chromatin in vitro: Enhancement of transcription by disruption of repressive nucleosome-nucleosome interactions. *The Journal of Biological Chemistry, 285*, 31954–31964.

Takahashi, T. S., Basu, A., Bermudez, V., Hurwitz, J., & Walter, J. C. (2008). Cdc7-Drf1 kinase links chromosome cohesion to the initiation of DNA replication in Xenopus egg extracts. *Genes and Development, 22*, 1894–1905.

Takahashi, K., & Yanagida, M. (2000). Cell cycle. Replication meets cohesion. *Science*, *289*, 735–736.

Takemoto, A., Kimura, K., Yanagisawa, J., Yokoyama, S., & Hanaoka, F. (2006). Negative regulation of condensin I by CK2-mediated phosphorylation. *EMBO Journal*, *25*, 5339–5348.

Takemoto, A., Maeshima, K., Ikehara, T., Yamaguchi, K., Murayama, A., Imamura, S., et al. (2009). The chromosomal association of condensin II is regulated by a noncatalytic function of PP2A. *Nature Structural and Molecular Biology*, *16*, 1302–1308.

Takemoto, A., Murayama, A., Katano, M., Urano, T., Furukawa, K., Yokoyama, S., et al. (2007). Analysis of the role of Aurora B on the chromosomal targeting of condensin I. *Nucleic Acids Research*, *35*, 2403–2412.

Tavormina, P. A., Come, M. G., Hudson, J. R., Mo, Y. Y., Beck, W. T., & Gorbsky, G. J. (2002). Rapid exchange of mammalian topoisomerase II alpha at kinetochores and chromosome arms in mitosis. *The Journal of Cell Biology*, *158*, 23–29.

Thiru, A., Nietlispach, D., Mott, H. R., Okuwaki, M., Lyon, D., Nielsen, P. R., et al. (2004). Structural basis of HP1/PXVXL motif peptide interactions and HP1 localisation to heterochromatin. *EMBO Journal*, *23*, 489–499.

Thoma, F., Koller, T., & Klug, A. (1979). Involvement of histone H1 in the organization of the nucleosome and of the salt-dependent superstructures of chromatin. *The Journal of Cell Biology*, *83*, 403–427.

Toyoda, Y., & Yanagida, M. (2006). Coordinated requirements of human topo II and cohesin for metaphase centromere alignment under Mad2-dependent spindle checkpoint surveillance. *Molecular Biology of the Cell*, *17*, 2287–2302.

Trinkle-Mulcahy, L., Andersen, J., Lam, Y. W., Moorhead, G., Mann, M., & Lamond, A. I. (2006). Repo-Man recruits PP1 gamma to chromatin and is essential for cell viability. *The Journal of Cell Biology*, *172*, 679–692.

Trinkle-Mulcahy, L., Andrews, P. D., Wickramasinghe, S., Sleeman, J., Prescott, A., Lam, Y. W., et al. (2003). Time-lapse imaging reveals dynamic relocalization of PP1gamma throughout the mammalian cell cycle. *Molecular Biology of the Cell*, *14*, 107–117.

Tsukahara, T., Tanno, Y., & Watanabe, Y. (2010). Phosphorylation of the CPC by Cdk1 promotes chromosome bi-orientation. *Nature*, *467*, 719–723.

Uchida, K. S., Takagaki, K., Kumada, K., Hirayama, Y., Noda, T., & Hirota, T. (2009). Kinetochore stretching inactivates the spindle assembly checkpoint. *The Journal of Cell Biology*, *184*, 383–390.

Uhlmann, F., Wernic, D., Poupart, M. A., Koonin, E. V., & Nasmyth, K. (2000). Cleavage of cohesin by the CD clan protease separin triggers anaphase in yeast. *Cell*, *103*, 375–386.

Vagnarelli, P. (2012). Mitotic chromosome condensation in vertebrates. *Experimental Cell Research*, *318*, 1435–1441.

Vagnarelli, P., & Earnshaw, W. C. (2004). Chromosomal passengers: The four-dimensional regulation of mitotic events. *Chromosoma*, *113*, 211–222.

Vagnarelli, P., & Earnshaw, W. C. (2012). Repo-Man-PP1: A link between chromatin remodelling and nuclear envelope reassembly. *The Nucleus*, *3*, 138–142.

Vagnarelli, P., Hudson, D. F., Ribeiro, S. A., Trinkle-Mulcahy, L., Spence, J. M., Lai, F., et al. (2006). Condensin and Repo-Man-PP1 co-operate in the regulation of chromosome architecture during mitosis. *Nature Cell Biology*, *8*, 1133–1142.

Vagnarelli, P., Morrison, C., Dodson, H., Sonoda, E., Takeda, S., & Earnshaw, W. C. (2004). Analysis of Scc1-deficient cells defines a key metaphase role of vertebrate cohesin in linking sister kinetochores. *EMBO Reports*, *5*, 167–171.

Vagnarelli, P., Ribeiro, S. A., & Earnshaw, W. C. (2008). Centromeres: Old tales and new tools. *FEBS Letters*, *582*, 1950–1959.

Vagnarelli, P., Ribeiro, S., Sennels, L., Sanchez-Pulido, L., de Lima Alves, F., Verheyen, T., et al. (2011). Repo-Man coordinates chromosomal reorganization with nuclear envelope reassembly during mitotic exit. *Developmental Cell, 21,* 328–342.

Valls, E., Sanchez-Molina, S., & Martinez-Balbas, M. A. (2005). Role of histone modifications in marking and activating genes through mitosis. *The Journal of Biological Chemistry, 280,* 42592–42600.

Varier, R. A., Outchkourov, N. S., de Graaf, P., van Schaik, F. M., Ensing, H. J., Wang, F., et al. (2010). A phospho/methyl switch at histone H3 regulates TFIID association with mitotic chromosomes. *EMBO Journal, 29,* 3967–3978.

Verdaasdonk, J. S., & Bloom, K. (2011). Centromeres: Unique chromatin structures that drive chromosome segregation. *Nature Reviews Molecular Cell Biology, 12,* 320–332.

Waizenegger, I. C., Hauf, S., Meinke, A., & Peters, J. M. (2000). Two distinct pathways remove mammalian cohesin from chromosome arms in prophase and from centromeres in anaphase. *Cell, 103,* 399–410.

Wang, F., Dai, J., Daum, J. R., Niedzialkowska, E., Banerjee, B., Stukenberg, P. T., et al. (2010). Histone H3 Thr-3 phosphorylation by Haspin positions Aurora B at centromeres in mitosis. *Science, 330,* 231–235.

Wang, F., Ulyanova, N. P., van der Waal, M. S., Patnaik, D., Lens, S. M., & Higgins, J. M. (2011). A positive feedback loop involving Haspin and Aurora B promotes CPC accumulation at centromeres in mitosis. *Current Biology, 21,* 1061–1069.

Warren, W. D., Steffensen, S., Lin, E., Coelho, P., Loupart, M., Cobbe, N., et al. (2000). The Drosophila RAD21 cohesin persists at the centromere region in mitosis. *Current Biology, 10,* 1463–1466.

Wheatley, S. P., Hinchcliffe, E. H., Glotzer, M., Hyman, A. A., Sluder, G., & Wang, Y. (1997). CDK1 inactivation regulates anaphase spindle dynamics and cytokinesis in vivo. *The Journal of Cell Biology, 138,* 385–393.

Wolf, F., Wandke, C., Isenberg, N., & Geley, S. (2006). Dose-dependent effects of stable cyclin B1 on progression through mitosis in human cells. *EMBO Journal, 25,* 2802–2813.

Woodcock, C. L., Frado, L. L., & Rattner, J. B. (1984). The higher-order structure of chromatin: Evidence for a helical ribbon arrangement. *The Journal of Cell Biology, 99,* 42–52.

Wurzenberger, C., Held, M., Lampson, M. A., Poser, I., Hyman, A. A., & Gerlich, D. W. (2012). Sds22 and Repo-Man stabilize chromosome segregation by counteracting Aurora B on anaphase kinetochores. *The Journal of Cell Biology, 198,* 173–183.

Xing, H., Vanderford, N. L., & Sarge, K. D. (2008). The TBP-PP2A mitotic complex bookmarks genes by preventing condensin action. *Nature Cell Biology, 10,* 1318–1323.

Xing, H., Wilkerson, D. C., Mayhew, C. N., Lubert, E. J., Skaggs, H. S., Goodson, M. L., et al. (2005). Mechanism of hsp70i gene bookmarking. *Science, 307,* 421–423.

Yamagishi, Y., Honda, T., Tanno, Y., & Watanabe, Y. (2010). Two histone marks establish the inner centromere and chromosome bi-orientation. *Science, 330,* 239–243.

Yamagishi, Y., Sakuno, T., Shimura, M., & Watanabe, Y. (2008). Heterochromatin links to centromeric protection by recruiting shugoshin. *Nature, 455,* 251–255.

Yeong, F. M., Hombauer, H., Wendt, K. S., Hirota, T., Mudrak, I., Mechtler, K., et al. (2003). Identification of a subunit of a novel Kleisin-beta/SMC complex as a potential substrate of protein phosphatase 2A. *Current Biology, 13,* 2058–2064.

Young, D. W., Hassan, M. Q., Pratap, J., Galindo, M., Zaidi, S. K., Lee, S. H., et al. (2007). Mitotic occupancy and lineage-specific transcriptional control of rRNA genes by Runx2. *Nature, 445,* 442–446.

Young, D. W., Hassan, M. Q., Yang, X. Q., Galindo, M., Javed, A., Zaidi, S. K., et al. (2007). Mitotic retention of gene expression patterns by the cell fate-determining transcription factor Runx2. *Proceedings of the National Academy of Sciences of the United States of America, 104,* 3189–3194.

Yue, Z., Carvalho, A., Xu, Z., Yuan, X., Cardinale, S., Ribeiro, S., et al. (2008). Deconstructing Survivin: Comprehensive genetic analysis of Survivin function by conditional knockout in a vertebrate cell line. *The Journal of Cell Biology, 183*, 279–296.

Zaidi, S. K., Young, D. W., Pockwinse, S. M., Javed, A., Lian, J. B., Stein, J. L., et al. (2003). Mitotic partitioning and selective reorganization of tissue-specific transcription factors in progeny cells. *Proceedings of the National Academy of Sciences of the United States of America, 100*, 14852–14857.

Zatsepina, O. V., Poliakov, V., & Chentsov Iu, S. (1983). Electron microscopic study of the chromonema and chromomeres in mitotic and interphase chromosomes. *Tsitologiia, 25*, 123–129.

Zeitlin, S. G., Shelby, R. D., & Sullivan, K. F. (2001). CENP-A is phosphorylated by Aurora B kinase and plays an unexpected role in completion of cytokinesis. *The Journal of Cell Biology, 155*, 1147–1157.

AUTHOR INDEX

Note: Page numbers followed by "*f*" indicate figures, and "*t*" indicate tables.

SUBJECT INDEX

Note: Page numbers followed by "*f*" indicate figures, and "*t*" indicate tables.

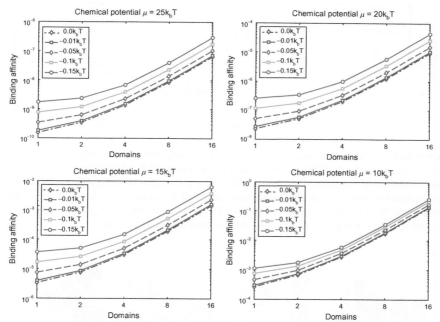

Christoph J. Feinauer *et al.*, Figure 3.8 The competitive result. Binding of several 16 bp spanning protein with different numbers of domains as indicated. All proteins were in the same system. The plot shows their binding affinity (absolute probability to find a DNA unit bound by a protein) averaged over 20 bp in the inner region of the DNA chain with length $N = 256$ bp. The binding energy per bound unit ε and the chemical potential μ were used as parameters as shown.

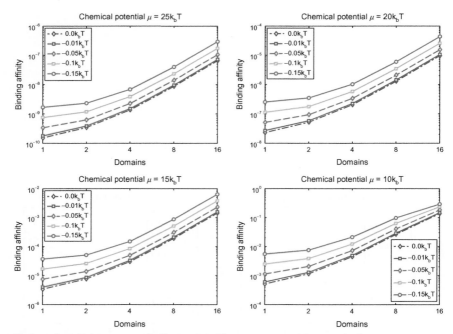

Christoph J. Feinauer *et al.*, Figure 3.9 The noncompetitive result. Separate systems were set up, each of which contained one protein type with a 16-bp binding interface and number of domains as indicated. The plot shows the binding affinity (absolute probability to find a DNA unit bound by a protein) averaged over 20 bp in the inner region of the DNA chain with length $N = 256$ bp. The binding energy per bound unit ε and the chemical potential μ were used as parameters as shown.

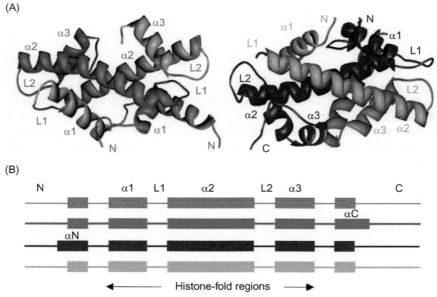

Amutha Ramaswamy and Ilya Ioshikhes, Figure 4.2 (A) The secondary structure of handshake motifs; H3–H4 and H2A–H2B formed by the histones H3 (pink), H4 (green), H2A.Z (blue), and H2B (orange) and (B) their schematic representation as colored in (A).

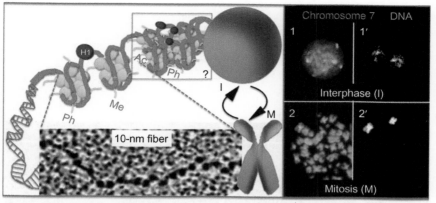

Paola Vagnarelli, Figure 6.1 Schematic representation of the organization of DNA into the 10-nm fiber: the first step of chromatin compaction. The histone tails within the nucleosome structure (green) can be posttranslational modified with phosphate groups (Ph), methyl groups (Me), or acetyl groups (Ac). How the higher levels of compactions are achieved is still debatable as indicated by the question mark in the gray box. The 10-nm fiber can be visualized as "beads on a string" by electron microscopy as shown in the EM image below the drawing. EM image kindly provided by Prof. W. C. Earnshaw, Edinburgh. The transition from interphase (I) to mitosis (M) is characterized by a remarkable change in the organization of the chromatin (compare 1 and 2) but the effective level of condensation is only two- to threefold variation between the two stages. Compare the area occupied by chromosome 7 (chromosome painting by FISH) in the interphase nucleus (1–1′) versus the area occupied in the mitotic cell (2–2′). (1–2) Chromosome painting with a chromosome 7-specific library of a human lymphoma cell line. FISH images kindly provided by Dr. Sabrina Tosi (Brunel).

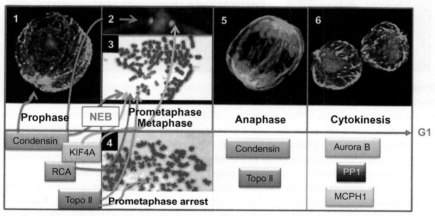

Paola Vagnarelli, Figure 6.2 Chromatin reorganization during mitosis requires several sequential steps that are regulated and coordinated by different chromosome-associated or regulatory proteins. 1: prophase cell (red: DNA, green: tubulin); 2: metaphase cells. A chromosome (red) is attached to the spindle microtubules (white) via the kinetochore (green). 3: Chromosome in prometaphase assume the characteristic x-shape structure and the sister chromatids are becoming visible. 4: Cell blocked in prometaphase by a spindle poison drug. The chromosomes maintain their mitotic structure while they undergo a further reduction in length. Note that the separation between sister chromatids is more pronounced than in (3) but they are still joined at the primary constriction. 4: Anaphase cell: the sister chromatids (red) are migrating toward the spindle poles and maintain their rod-shape structure. (green: tubulin). 6: Telophase-cytokinesis: the chromosomes reach their maximum compaction in mitosis. See text for explanations on the role of different proteins.

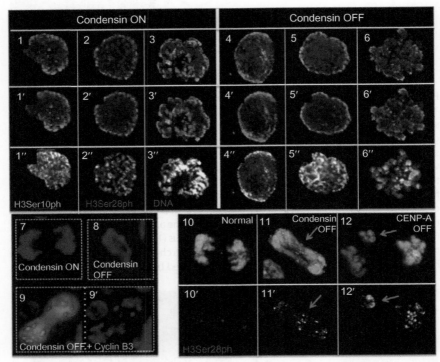

Paola Vagnarelli, Figure 6.5 1–6: Stages of mitotic chromosome condensation and some histone modifications that are associated with the process. 1–3 normal cells: signs of chromosome condensation are already visible at the first appearance of phosphorylation on histone H3 at Ser28 (1–1″), progresses further with recognizable chromosome formation when phosphorylation on histone H3 at Ser28 is completed. In cells without condensin, there is a delay in the earliest stages of chromosome condensation with a clear sign of condensin structures only at the nuclear envelope breakdown (5–5′). 7–9: Cells lacking condensin present abnormal segregation in anaphase with the appearance of chromatin bridges that persist in late anaphase/telophase (8). This phenotype is alleviated by expressing high levels of CyclinB3 in anaphase (green in 9); compare the anaphase chromatin in 8 with the one in 9′. 10–12: Chromatin that does not segregate correctly in anaphase maintains a high level of phosphorylation on histone H3 (11, 11′; 12, 12′) and this could represent the signal for blocking cell division. See text for discussion.